MULTICULTURAL MATH

Hands-On Math Activities from Around the World

Claudia Zaslavsky

SCHOLASTIC
PROFESSIONAL BOOKS

New York • Toronto • London • Auckland • Sydney

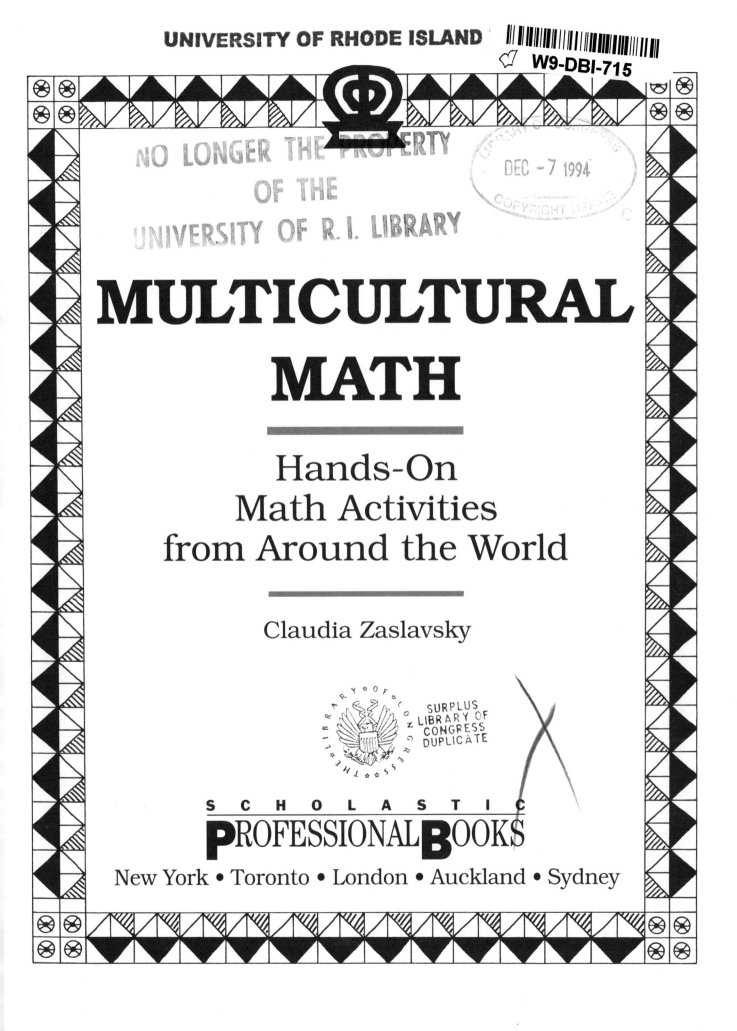

To Clara, David, Sarah,
and all the children
in the world

Cover design by Vincent Ceci
Cover illustration Marti Shohet
Interior design by Roberto Dominguez and Jaime Lucero
Interior illustrations by Mona Mark
ISBN 0-590-49646-8
Copyright ©1994 by Claudia Zaslavsky. All rights reserved.
12 11 10 9 8 7 6 5 4 3 2 1 1 2 3 4 5 / 9
Printed in the U.S.A.

TABLE OF CONTENTS

Part III: Space, Shape, and Size

Part IV: Fun and Games

Introduction

This book of multicultural mathematical lessons is appropriate for all students. It offers insights into the development of mathematical concepts in many societies of the past and the present.

Many teachers are introducing multicultural perspectives in their social studies, language arts, music, and art lessons. They know that all peoples have contributed to history, literature, and the arts, and that they all have important stories to tell. But multicultural perspectives in math education? Isn't math the same all over the world? Isn't two plus two always four?

It's true that people all over the world engage in similar mathematical activities. They count objects, they measure distances and other quantities, they design buildings and works of art, they play games that involve mathematical ideas. But how they carry out these activities and how they talk about them differs from one society to another. Inclusion of such ideas can enrich the mathematics curriculum in many ways.

According to the *Curriculum and Evaluation Standards for School Mathematics* of the National Council of Teachers of Mathematics (1989): "Students should have numerous and varied experiences related to the cultural, historical, and scientific evolution of mathematics." Students learn that mathematics arose out of the real needs and interests of society. They are challenged to solve the problems that face people today and that confronted them in the past. Multicultural mathematics engages students' imagination and helps them to develop skills in critical thinking and analysis that can be applied to all areas of life.

How this Book Is Organized

The book is divided into four main sections: Numbers and Numerations Systems; Using Numbers in Real Life; Space, Shape, and Size; Fun and Games. Teacher resource material with each lesson provides a description of the activities, background information, answers to problems, and suggestions for further activities and research. Relevant literature for students is found at the beginning of each section. Each lesson contains two reproducible sections of activities and problems for students. The second section is more advanced than the first. They may be used in sequence or independently. Each section includes background information and proposals for activities that encourage cooperation, creativity, and critical thinking. The variety of activities should appeal to a diversity of learning styles and backgrounds. Many are open-ended to engage students of varied interests and ability levels.

Mathematical Concepts and Connections Strands recommended in the NCTM *Standards* (1989) for grades K-4 and 5-8 are included in the lessons. Emphasis is placed on the problem-solving process and on communication, reasoning, and connections, as well as on the more specific mathematical skills, including the use of calculators and appropriate manipulatives.

The introduction of multicultural perspectives encourages connections among mathematical topics and other subject areas. For example, the activities dealing with home-building involve measurement, geometry, computation, estimation, and problem solving. Students must make decisions about size and shape, building materials, durability, and many other questions. They read relevant literature and learn about the influence of traditions. They use their artistic abilities to plan, build, and decorate model homes. They discuss and write about these ideas, enhancing their skills in communication. They observe architecture in their own environment, and consult with their families and with community members to learn their experiences. As they interact with their environment and study the mathematical ideas of other societies, they develop skills in analysis that will serve them throughout their lives.

 ## How to Use this Book

Teachers in the elementary grades have the advantage of teaching all or most of the subjects. They can incorporate mathematics into a social studies lesson about Native Americans or discuss a book like *The Village of Round and Square Houses* (Grifalconi, 1986) in connection with a lesson on the area and perimeter of round houses. Mathematics teachers might consult with the relevant teachers of other subjects about integrating math with those subjects.

Most of the activities can best be carried out by students working in pairs or in cooperative learning groups. (Working in pairs is essential for the games in Part IV.) You can further the goals of multicultural education by placing in the same group students of varied abilities and ethnic/racial backgrounds.

You may want to begin the lesson with a general introduction to the topic, asking students to contribute what they know about it and engaging their interest. Encourage students to check their own or each other's work, rather than looking to you as the authority. Accept all solutions without comment, whether favorable or adverse, but expect students to justify their answers. Help them to gain confidence in their ability to control their own learning.

As you walk around the room while students are working, you will hear their conversation. How are they approaching the problems in the lesson? Encourage them to share their ideas with the group and the class. Be sure that they take the time to reflect upon the mathematical concepts they are using. Students may become so involved in the activity that they neglect to reflect upon the mathematical concepts in the lesson. By throwing out challenging questions and fostering discussion, you can make sure that this doesn't happen. Every lesson includes some form of written or artistic work on paper. Students can collect them into portfolios that can form the basis for their assessment.

Some teachers and administrators are concerned that time spent on projects such as those described in this book is time away from drill for standardized tests. But research has shown conclusively that when students are motivated to do math, when they are involved in exciting mathematical activities rather than devoting large chunks of time to drilling for the tests, they score just as well in computation and score far higher in problem solving. Newer tests do include problem-solving types of questions, and these activities can prepare students to handle them.

 ## Multiculturalism

According to the NCTM *Standards* (1989), "students' cultural backgrounds should be integrated into the learning experience." Children take pride in their culture when it is recognized in the classroom. At the same time, we cannot assume that every child of a specific ancestry knows about or is interested in the heritage and culture of that society. We should be sensitive to students' relationship to their heritage.

Many school districts are encouraging, even mandating, the inclusion of a multicultural perspective in all subject areas. Publishers are responding by making available a variety of materials. No teacher has the background to evaluate the authenticity of all these materials. What can a teacher do to ensure that students receive accurate information?

Here are a few guidelines:
- A culture should not be portrayed as "exotic" because it is different from that of the writer or publisher.
- Africa should not be portrayed as though it is a country, rather than a continent where people speak a thousand different languages. Some publishers devote separate books to Mexico, Israel, and many other countries, but have just one book for the whole continent of Africa.Steer away from these books.
- A culture should be portrayed accurately. This is a difficult problem, but Doris Seale and Beverly Slapin come to the rescue with their 1992 book, *Through Indian Eyes: The Native Experience in Books for Children*. Their guidelines for evaluating children's literature apply also to other cultures, and their analysis of specific books provides valuable insights. Many children have stereotypical ideas about Native peoples, and we should do everything possible to correct these false impressions.

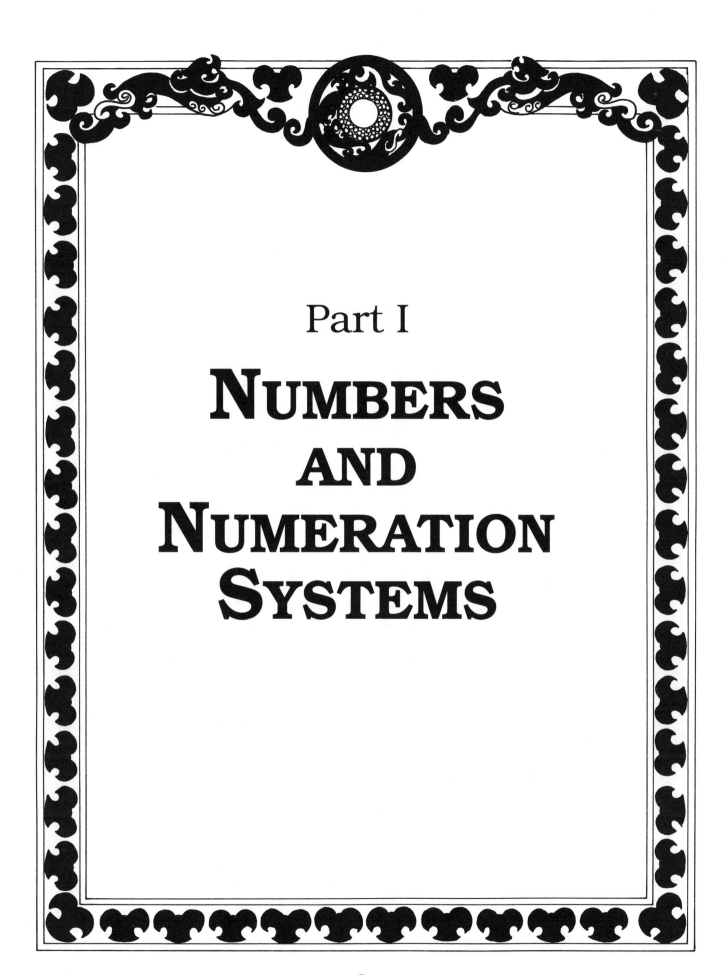

Part I

NUMBERS AND NUMERATION SYSTEMS

The six lessons in this section deal with the names of numbers and the ways in which numbers are expressed in various societies.

Lesson 1 introduces counting on the fingers, among the Plains Indians and the Kamba people of Kenya.

Lessons 2 and 3 deal with names for numbers. Your students will learn that as people find that they need large numbers, they develop *systems* of numeration that facilitate the expression of these quantities, based on grouping. Number words in English are based on groups of ten and powers of ten—hundreds, thousands, etc. Some numeration systems group by fives and twenties, while others use both ten and twenty as bases.

Lessons 4 and 5 deal with several ancient systems of written numerals: Egyptian, Chinese rod, and Mayan. Although their number words differ, most people today use the familiar Indo-Arabic numerals that originated in India about 14 centuries ago. The system has two outstanding features: ten digits (including zero) and place value. It was transmitted by the Arabs to other parts of Asia, to north Africa, and to southern Europe. It was not until centuries later that the system became popular in Europe and began to replace Roman numerals. Both the Chinese and the Egyptian systems have ten as the base. The Maya system, based on five and twenty, also included one of the earliest symbols for zero.

Lesson 6 introduces the Russian abacus and the Chinese abacus, calculating devices that are still in use today.

Book Links for Students

Carona, Philip. *Numbers*. Chicago: Children's Press, 1982.

Dilson, Jesse. *The Abacus: A Pocket Computer*. New York: St. Martin's Press, 1968.

Feelings, Muriel. *Moja Means One: Swahili Counting Book*. New York: Dial Press, 1971.

Haskins, Jim. *Count Your Way Through . . .* series. Minneapolis: Carolrhoda Books.

Leaf, Margaret. *Eyes of the Dragon*. New York: Lothrop, Lee & Shepard, 1987.

Lumpkin, Beatrice. *Senefer: A Young Genius in Old Egypt*. Trenton, NJ: Africa World Press, 1992.

Wahl, John, and Stacy Wahl. *I Can Count the Petals of a Flower*. Reston, VA: National Council of Teachers of Mathematics, 1976.

Zaslavsky, Claudia. *Count on Your Fingers African Style*. New York: Crowell, 1980.

———. Series on science in ancient cultures by Franklin Watts, 1988.

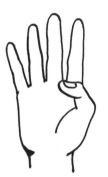

COUNTING ON FINGERS

Background

In many languages the names for numbers are related to counting on the fingers or on other parts of the body. In English, for example, the word "eleven" means one left after counting all ten fingers. Indeed, grouping by tens, called "base ten," probably originated with finger counting.

Getting Started

Ask students whether they have ever used their fingers when counting. Make a class list of the various times. Then, ask students to classify the list in some way. Perhaps they counted the number of days (weeks, months, etc.) before an event took place. They may have counted how many guests to invite. Did they use their fingers to do math calculations? What words did they say as they ticked off the numbers on their fingers?

Invite students to imagine that they are in a market in a foreign land. Ask: "You see some nice bananas. How can you show that you want three bananas?" They will suggest using the fingers. "Think how you will show three on your fingers, but don't do it until I say 'Now.' Will you use one hand or two? Which fingers will you use? Will you raise them or bend them down?"

Students should hold up their hands showing their methods, while they look at their classmates' methods. After they have lowered their hands, ask questions like: "Who used two hands? Who used the right hand alone? The left hand alone?

OBJECTIVES FOR STUDENTS

♦ To connect the number names with the finger gestures.

♦ To analyze systems of finger gestures.

♦ To use finger gestures in every-day situations.

♦ To invent a system of finger gestures.

Who used the thumb?" and so forth. Children will see that there are many ways to solve this simple problem.

Carry out a similar exercise with the number eight. Then you might have the class read Zaslavsky's *Count on Your Fingers African Style.* Children will learn that different peoples have specific ways of counting on their fingers.

RIGHT OR LEFT?

The Zulu system of finger counting is identical to that of the Plains Indians, except that Zulus start with the left hand and Plains Indians with the right.

Extension Activities

1. Ask students to discuss the Kamba finger-counting system, then write a description of it so that a reader would understand the system without seeing the pictures.

2. The Mende people live in Sierra Leone, a country in West Africa. Their word for twenty is *nu gboyongo.* It means "a whole person." Ask students to think of a reason.

3. Have students work in groups to plan a scene in a Kamba market in which buyers and sellers use Kamba number words and finger gestures.

4. Students can investigate the gesture systems of other cultures.

5. Ask students to work in groups to make up their own finger-counting system from one to ten.

6. Find a book on Chisanbop and teach the children this system of finger reckoning, based on the Korean and Japanese abacus.

Answers to Student Pages

Page 14:
1) 2: raise the right index and middle fingers; 4: raise all four fingers of the right hand; 7: hold the small and ring fingers of the left hand; 9: hold four fingers of the left hand.

Name: _____

 # INDIANS OF THE GREAT PLAINS

It's handy to have some objects to help you count. What objects can be handier than your fingers?

Some people have special ways of using their fingers when they count. They may start on the right hand or on the left. They may begin with the little finger or with the thumb, or maybe with the index finger.

When people who speak different languages come together, they invent a sign language understood by many groups. That is what happened hundreds of years ago when many different tribes of American Indians lived on the Great Plains. They made up the signs below for the numbers from one to ten.

With a partner, practice the number signs below. Try guessing each other's numbers without looking at the diagram.

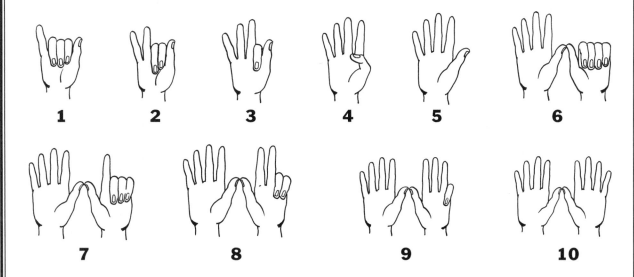

Now Try This

☞ Imagine you are talking to a friend on the phone and you want to describe the gestures used for the Plains Indians' finger-counting system. How would you describe the system to your friend? Write that description on the back of this sheet and share it with the class.

Name: _____

THE KAMBA SYSTEM OF KENYA

Many people use finger gestures to show numbers. But just as people speak different languages, they have different systems of finger counting, too.

The Kamba people of Kenya, in East Africa, have a special way of using their fingers when counting. Here are some of their counting words and gestures.

Draw a picture for each missing sign below.

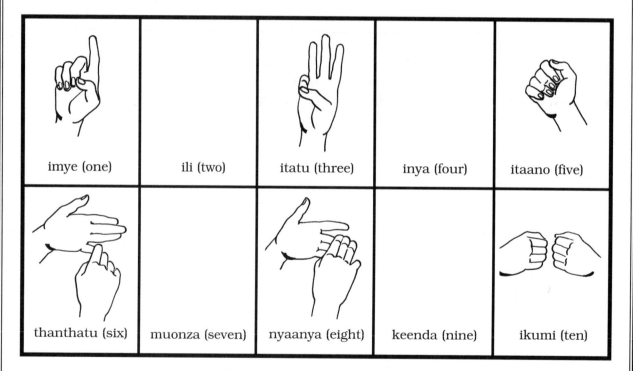

| imye (one) | ili (two) | itatu (three) | inya (four) | itaano (five) |
| thanthatu (six) | muonza (seven) | nyaanya (eight) | keenda (nine) | ikumi (ten) |

Now Try This ━━━

☞ Practice the Kamba signs and number words with a partner. Show a sign and ask your friend to say the Kamba number. Then, do the opposite: Say a number in English and ask your partner to say the Kamba number and show the sign. Take turns.

NAMES FOR NUMBERS

Background

This lesson, as well as the next, deals with number *names*, not with the written symbols called numerals. Be careful that students do not confuse the two concepts. See the discussion in the introduction to Part I (page 10).

Getting Started

Ask students how they use number words in everyday life and make a list of these applications. Challenge them to find substitutes for the number words in a specific instance, such as "I need *some* money for lunch." Does the substitute term give adequate information? Can we manage without numbers?

Some students may be able to count in a language other than English. Make them proud of their skill by asking them to share it with the class. You might invite family members who can count in foreign languages to share their knowledge with the class.

Wall maps of Europe, Africa, and the Americas will be helpful for this lesson. Ask your students to locate on the map the regions in which these languages are spoken. Spanish, for example, has a wide range in the Americas.

Extension Activities

1. Challenge students to find reasons for the similarity of the number words in some different languages.

OBJECTIVES FOR STUDENTS

♦ To compare the number words in several languages.

♦ To gain a better understanding of the concept of base by analyzing the structure of number systems.

2. Encourage students to memorize the counting words in specific languages. A group might put on a skit in which they use the number names in that language and challenge the class to guess the meaning.

3. Ask students how they could use pennies, nickels, and dimes to show Kpelle numbers and groups? (Pennies, nickels, and dimes embody the grouping system of the Kpelle number words.) What other manipulatives can they use?

Answers to Student Pages

Page 17:
1) Fourteen, eighteen, twelve, twenty-one. 2) Neunzehn, diecinueve, dicianove. The Italian word is probably a surprise because the word for ten precedes nine, unlike the order in sedici (sixteen).

Page 18:
1) Seven, eight, fourteen, eighteen, twenty-two, thirty-six. Note: noolu mei feere means "five over two," or two left over after counting five fingers on a hand. Now Try This) Kpelle number words involve grouping by fives and by tens, related to counting on one hand and then on the other hand.

Name: _____

 # GERMAN, SPANISH, AND ITALIAN

Some people can count in other languages. Can you? Here are lists of counting words in English, German, Spanish, and Italian.

English	*German*	*Spanish*	*Italian*
one	eins (eye'ns)	uno (oo-noh)	uno
two	zwei (ts'vy)	dos	due (doo-eh)
three	drei (dry)	tres	tre (treh)
four	vier (feer)	cuatro	quattro
five	funf (finf)	cinco	cinque (cheen-kweh)
six	sechs (seks)	seis (seh-iss)	sei
seven	sieben	siete (syeh-teh)	sette (seh-teh)
eight	acht (aht)	ocho	otto
nine	neun (noin)	nueve (n'weh-veh)	nove (noh-veh)
ten	zehn (tsehn)	diez (d'yess)	dieci (d'yeh-ch´ée)
eleven	elf	once (ohn-seh)	undici (oon-deechee)
sixteen	sechzehn	dieciseis (d'yes-ee-seh-is)	sedici (seh-deechee)
twenty	zwanzig (ts' vahn-tsik)	veinte (veh-een-teh)	venti

1. What do you think is the English number name for:

vierzehn (German) _____   dieciocho (Spanish) _____

dodici (Italian) _____   veintiuno (Spanish) _____

2. On a separate sheet guess the word for nineteen in each language.

Now Try This

☞ Compare all the words for each number. How are they alike? How are they different? Are some words more alike than others? For example, four and vier are almost alike, and so are cuatro and quattro. Why do you think that is so?

Name: _____

 # THE KPELLE OF LIBERIA

About a thousand different languages are spoken on the continent of Africa. Here's how the Kpelle people, who live in Liberia (in West Africa), count.

Here are some Kpelle number words:

English	Kpelle	English	Kpelle
one	taa	six	noolu mei taa
two	feere	ten	buu
three	saaba	twelve	buu kau feere
four	naang	twenty	buu feere
five	noolu	twenty-five	buu feere kau loolu

Note: The word mei means "over" and kau means "and." The letter "n" can become the letter "l".

Try to figure out the English names for these Kpelle counting words:

noolu mei feere _____   noolu mei saaba _____
buu kau naang _____   buu kau loolu mei saaba _____
buu feere kau feere _____
buu saaba kau loolu mei taa _____

Now Try This

☞ English number words are based on groups of ten. For example, "fourteen" means four plus ten, and "forty" means four times ten. Ten is the base of the counting system in English. How are groups formed in the Kpelle counting system? On a separate sheet write all you know about Kpelle numbers.

MORE NAMES FOR NUMBERS

Getting Started

Ask students to give examples of words that imply specific numbers of objects. They might offer such words as dozen, score, pair. You may have to give them hints. For example, ask them to think of babies (triplets, quintuplets, etc.), musicians (duo, quartet), years (decade, century), computer memory (kilobyte, megabyte), etc.

Also, ask students the meaning of the words forty and eighty (four tens, eight tens).

Extension Activities

OBJECTIVES FOR STUDENTS

◆ To learn how numbers are named in French, Mayan, Yupik (Alaska), and Igbo (Nigeria).

◆ To better understand the concept of base by analyzing several systems that group by twenties.

1. Ask a student to read Lincoln's words (on page 21) aloud. You might have the class analyze the phrase "all men are created equal." Note that women and slaves were not allowed to vote. By amendments to the Constitution, African-American men received the right to vote after the Civil War, while women had to wait until 1920 for this right. The class may want to discuss some of the *inequities* that still exist today.

2. Ask students to locate France on the map. Then ask students to locate French-speaking countries and regions, such as Quebec in Canada; Mali, Senegal, Zaire, etc. in Africa; former colonies in Asia; Haiti and several Caribbean islands. Some students might need your assistance.

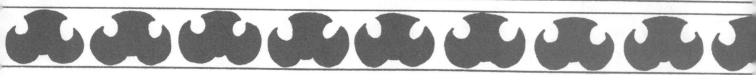

3. Ask students to locate the regions inhabited by the Maya, the Yupik, and the Igbo.

4. Ask students to devise materials to represent a base-20 system. They might think in terms of units, fives, and twenties to represent individual digits, a hand or foot, and a whole person. Have students make up addition and subtraction exercises using these materials.

5. Challenge students to find reasons why peoples develop systems based on grouping by tens or twenties.

> ## WHERE ARE THEY NOW?
> About two million Maya still live in southern Mexico and northern Central America. The Yupik Eskimos live in western Alaska. Their lifestyle is quite different from that of the more northern Inuit Eskimos; for example, they never built igloos. The Igbo of southeastern Nigeria, both men and women, have long been known as sophisticated traders.

Answers to Student Pages

Page 21:
1) 1776, the year in which the Declaration of Independence proclaimed the independence of the American colonies from Great Britain. 2) Four tens, or four times ten. 3) Sixty plus ten, or seventy; four times twenty, or eighty; four times twenty plus ten, or ninety. Now Try This) The French language is interesting because some number words depend upon grouping by tens, while others group by twenties.

Page 22:
1) ox kal; yuinaat pingayun; ohu ato. 2) Mayan: two hundred, eighty.
Yupik: eighty, sixty. Igbo: two hundred, eighty. 3) 146 = 7 x 20 + 6;
235 = 11 x 20 + 15;   399 = 19 x 20 + 19.

> ## SIMILAR SYSTEMS
> The French, the Maya, the Yupik, and the Igbo peoples live in different parts of the world yet their number systems have a common feature: They're based on grouping by twenties.

Name: _____

 # ENGLISH AND FRENCH

Four score and seven years ago our fathers brought forth on this continent a new nation, conceived in liberty, and dedicated to the proposition that all men are created equal.

These words are famous. They are the beginning of the speech by President Abraham Lincoln at the dedication of the cemetery at Gettysburg, Pennsylvania for those who died in the Civil War. Here, Lincoln uses numbers in a way that was once quite common in English. As you'll see, the French still use it. Use the back of this sheet to answer the questions below.

1. A score means twenty. Lincoln gave his Gettysburg address in 1863. What year was four score and seven years before that date? What happened in that year?

2. Guess the meaning of the word *quarante*.

English	French	English	French
one	un (ang)	ten	dix (deess)
two	deux (duh)	twenty	vingt (vang)
four	quatre (kat'r)	forty	quarante (kar-AHNT)
six	six (seess)	sixty	soixante (swah-SAHNT)

The word soixante means six tens, or six times ten.

3. Here are more French number names. Guess their meaning in English.

soixante-dix (add soixante and dix)_____

quatre-vingt (multiply quatre and vingt) _____

quatre-vingt-dix _____

Now Try This ━━━━━━━━━━━━━━━━━━━━━

☞ Discuss how numbers are grouped in the French language.

Name: _____

MAYA, YUPIK (ALASKA), AND IGBO (NIGERIA)

Long before Columbus sailed to America, the Maya were living in southern Mexico and northern Central America. The Yupik Eskimos live in western Alaska. Igbo people live in eastern Nigeria. Even though these three groups live far apart, their number words have a common feature. They are all based on grouping by twenties.

English	Mayan (Yucatec)	Yupik	Igbo
one	hun	atauciq	otu
two	ca	malruk	abuo
three	ox	pingayun	ato
four	can	cetaman	ano
ten	lahun	qula	iri
twenty	hun kal	yuinaq	ohu
forty	ca kal	yuinaak malruk	ohu abuo

1. Guess the words for sixty in each language.

Mayan: _____   Yupik: _____

Igbo: _____

2. Work out what these names mean and write them in English.

Mayan: lahun kal _____   can kal _____

Yupik: yuinaat cetaman _____   yuinaat pingayun _____

Igbo: ohu iri _____   ohu ano _____

3. Write each number as groups of twenties plus a number less than twenty. Example: 94 = 4 x 20 + 14.

146 _____   235 _____   399 _____.

Now Try This ━━━━━━━━━━━━━━━━━━━━━━━

☞ Using activity 3 as a model, make up some examples for a classmate to complete.

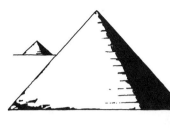

SYMBOLS FOR NUMBERS

Getting Started

Students may not know the origin and history of Indo-Arabic numerals. See the discussion in the introduction to Part I (page 10).

Ask students to locate Egypt on a map of Africa. The territory of Ancient Egypt extended into the present land of Sudan.

Students might make the symbols for 1, 10, and 100 from pipe cleaners or other pliable material. Manipulating the actual symbols can facilitate addition and subtraction, and clarify the operation of regrouping.

OBJECTIVES FOR STUDENTS

♦ To learn to use Egyptian hieroglyphic numerals.

♦ To compare the Egyptian numeration system with the Indo-Arabic system.

Extension Activities

1. Students will enjoy reading Lumpkin's *Senefer: A Young Genius in Old Egypt* (see bibliography on page 10), the fascinating story, based on historical events, of a boy who grows up to be a famous scientist; it includes Egyptian hieroglyphic numbers, addition, and multiplication by doubling.

2. Students may want to compare the Egyptian and Indo-Arabic systems of numerals listing all the ways they are alike and all the ways they are different.

3. Ask students to compare Egyptian and Roman numerals. Students should note that Roman numerals involve multiples of 5 as well as 10 (V=5, L=50, D=500) and also that place can make a difference: thus IX=9 while XI=11, XL=40 while LX=60, etc.

4. Ask students to work in groups to make up a story that might be written on a temple wall. They can write the words in English and include at least five Egyptian numerals. Remind students that the Egyptians wrote from right to left. Challenge groups to write their stories this way.

5. Challenge students to research the evolution of the Indo-Arabic numerals that we now use, or the ancient Greek and Hebrew numerals based on letters of their alphabets.

Answers to Student Pages

Page 25:
1) 1; 10; 12; 23; 46; 81.

2) ⦀⦀; ⦀⦀∩; ∩∩∩; ‖∩∩∩

The symbols may be strung out in a line or grouped to facilitate counting.

Page 26:
1) 612; 250; 1,306; 3,020.

2) ⦀∩999; ⦀∩∩⌇; 999⌇⌇

3. a. ⦀∩99 = 215;   b. ‖∩∩999 = 522

Name: _____

 # NUMBERS IN ANCIENT EGYPT

About 5,000 years ago Egypt became a powerful country. Large temples were built for the king, called the pharaoh (FAIR-oh). On the walls the builders wrote about the pharaoh and great events. They carved the words and numbers into the stone. Below are some examples of their numbers and what they mean.

|||	∩∩∩∩	''''''∩∩∩
3	40	39

1. Can you write these Egyptian numbers our way?

I	∩	II∩	IIII∩∩	''' ∩∩∩∩	I ∩∩∩∩ / ∩∩∩∩

2. On a separate sheet, write these numbers the Egyptian way:
   8, 14, 30, 62

3. Compare the numerals  **and 98.**

Which is easier to write, Egyptian or Indo-Arabic? Which is easier to understand? Explain your answers on a separate sheet.

Now Try This

☞ Write a story that has at least four different numbers in it. Write the numbers the Egyptian way.

Name: _____

 # HOW THE EGYPTIANS WORKED WITH NUMBERS

The ancient Egyptians carved the history of their victories and achievements on the stone walls and columns of their temples and pyramids. Their words and numbers began as pictures known as hieroglyphs (hy-roh-GLIFS).

1	10	100	1,000	10,000	1,238

1. Translate each Egyptian numeral into an Indo-Arabic numeral.

_____ _____ _____ _____

2. On a separate sheet, use Egyptian hieroglyphs to represent the numbers 317; 1,029; 2,300.

3. Add the Egyptian numerals, then check in Indo-Arabic numerals. Show all the steps in your work.

Example:

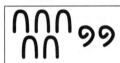

53
+37
90

a)

b)

Now Try This

☞ Write a subtraction problem in Egyptian numerals. Ask another student to solve and check it. Does it require regrouping?

SYMBOLS FOR NUMBERS

Background

Chinese rod numerals were in use at least 2,000 years ago, and were adopted by the Japanese about 14 centuries ago. The rods were placed on a counting board with columns for the ones, the tens, etc. An empty space stood for zero. Red rods represented positive numbers and black rods, negative numbers. People moved the sticks to add and subtract numbers.

The numeration system that we attribute to the Maya was already in use by earlier peoples in Middle America perhaps 3,000 years ago. The Maya flourished in the period from about 18 to 11 centuries before the present. Although they had extensive written documents, very few survived destruction by Spanish conquerors and priests, who considered them "works of the devil." Only in the last few years have the surviving documents and writings on stone tablets been adequately translated. These writings show a remarkable knowledge of astronomy and skill in calculation. The Maya were probably the first people to invent a symbol for zero.

> ### OBJECTIVES FOR STUDENTS
>
> ◆ To learn and use Chinese stick (or rod) numerals.
>
> ◆ To learn and use Mayan numerals.
>
> ◆ To compare these two systems with our standard Indo-Arabic numerals.

Getting Started

Students will need toothpicks and beans to represent numerals in this lesson. Have students locate China and the Mayan region on a map. Although it is difficult for children to conceptualize time, they may be impressed by the age of these systems. When studying the Chinese rod numerals, be sure students understand that, except for the rod representing five, the rods in the ones and hundreds columns are vertical, while those in the tens and thousands columns are horizontal. Share with them information about the cultures of these peoples.

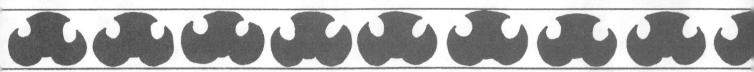

Extension Activities

1. Ask students to compare the Chinese system with ours. (Both systems have place value and grouping by tens (base ten), with the smallest value on the right. Both use different symbols for each digit from one to ten. The number of rods corresponds to the value of the number, except that a single rod represents a group of five. The Chinese system had two sets of symbols, which alternate from one column to the next. The Chinese had no symbol for zero.)

2. When the Maya bought and sold, they calculated with sticks and pebbles to represent the bars and dots of their numerals. Ask students to work in groups and make up skits about buying and selling, using Mayan numerals. Suggest using toothpicks and beans, or other materials to represent bars and dots.

3. Students can discuss the ways in which the Mayan numerals are like ours and the ways in which they are different.

Answers to Student Pages

Page 29:

1) **3** ⦀⦀ **7** ⫼ **=** **40** ≡ ⊥ **80**

2) 437;   2,910

3)

	1000	100	10	1
523		⦀⦀⦀	=	|||
1,064	—		⊥	||||
6,370	⊥	|||	⊥	

Page 30:

1) **4** •• **10** ⟂ **13** **16** ≣

2) 80; 147; 202; 315; 399.

Now Try This)

a. ≣ + — = ≣   b. ⟂ + •••• = ⟂

c. • + — = ⟂   d. ••• + ⟲ = •••   e. ⟂ + •• = ≣

28

Name: _____

 # CHINESE STICK NUMERALS

Here is a puzzle. Can you read this numeral? |||=TTT
Here are some hints:
- It has three digits.
- It has place value, just like the Indo-Arabic numerals we use.
- Think of using one hand or two hands to show each digit.

Did you guess that the numeral is 328? It is called a Chinese stick numeral.
This numeral is 1,792: —TT ≡|| Can you figure out the system?

1. Here are some of the numerals. Follow the pattern and fill in the missing numerals in the Chinese system and in our system.

1	2	3	4	5	6	7	8	9	
I	II	III		IIIII	T	TT	TTT		
10	20	30		50	60	70		90	100
—		≡	≣		⊥		≟	≡	I

2. Write each number in our system:

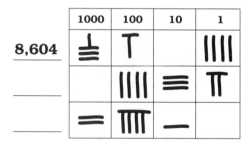

8,604

_____

3. Write the numbers below in Chinese stick numerals.

	1000	100	10	1
523				
1,064				
6,370				

Now Try This

☞ Work with a partner. Pretend that one of you is a merchant and the other is a customer. Make up a skit in which you add and subtract using Chinese stick numerals. Make a large counting board like the one in activity 2 and 3 above. Use toothpicks or other materials to form the numerals.

Name: _____

 # MAYAN NUMERALS

For thousands of years the Maya have lived in Mexico and Central America, where they built great temples and large cities. Using place value and just three symbols—bar, dot, and zero symbol—they were able to write numbers in the millions. Below are some of their symbols.

0	3	6	9	14	15	17	19

1. Write the missing Mayan or standard (Indo-Arabic) numerals.

	7		11			18

2. In a previous lesson, you learned some Mayan number words. The word for forty, ca kal, means "two times twenty." Study the examples, noting how the symbols are placed. Then write the meaning of each Mayan numeral in standard form. The shell is a symbol for zero.

69	232					

Now Try This

☞ Write the following sums in Mayan numerals. Show how you regroup when you add. Compare your work with a partner.

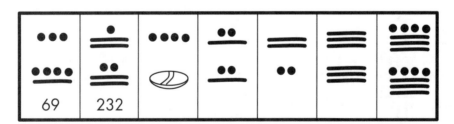

Example:
17 + 9 = ➡ ➡ = 26

    a. 13 + 5   b. 12 + 4   c. 25 + 5   d. 108 + 60   e. 240 + 87

30

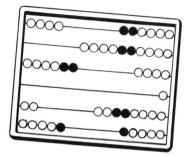

THE ABACUS

Background

When Napoleon invaded Russia in 1812, a mathematician traveling with the French army saw Russian people using an abacus called a *scety* (s-CHAW-tee). He thought it such a wonderful device for teaching arithmetic that he introduced the abacus into France. From there it spread to other European countries and to America.

Counting boards have been in use for thousands of years; Chinese rod numerals are a good example. The abacus offers the convenience of movable beads affixed to a counting board. The Chinese abacus, or *suan pan*, dates back about eight centuries. Years later the Japanese adopted it and reduced the number of beads on each rod so that there are only four on the lower level and one on the upper level.

OBJECTIVES FOR STUDENTS

◆ To learn about the abacus as a calculating device.

◆ To become familiar with and make a model of a Russian and a Chinese abacus.

◆ To become familiar with decimals and large numbers.

Getting Started

Have suitable materials available for students to choose from when making their abacuses. Pieces of heavy cardboard, strong thread or wire, beads or macaroni are some of the possibilities.

Discuss the first abacus drawing on the reproducible to make sure students understand how it works. (This applies to both reproducibles.)

Extension Activities

1. Students can make their own abacus. You might suggest that they string beads or macaroni on lengths of cord and staple them to stiff cardboard. Some students may even want to write an instruction manual to go with theirs.

2. Suggest that students learn about the Japanese abacus.

3. Have students research the Chisanbop (also called Fingermath) method of calculating on the fingers. This method, based on the Korean (similar to the Japanese) abacus, was introduced into the United States by Hang Young Pai, who has written several books about the method.

Answers to Student Pages

Page 33:
1) 297.45; 980.30
2)

Page 34:
1) 12,059; 53,680
2)

Now Try This) The suan pan is based on grouping by fives and tens. It is positional; each rod has a value ten times the value of the rod to its right. To record a number, push up the beads on the lower section one by one until five have been moved. Then exchange the five lower beads for one upper bead. Pull back the five lower beads and add them one by one to the upper bead until all five have been pushed up. Exchange these five for the second upper bead, indicating a total of ten. Exchange the ten for one bead on the rod to its left. Chinese rod numerals also depend upon grouping by fives and tens and involve place values.

Name: _____

 # THE RUSSIAN ABACUS

You may have seen a counting board that looks like this: It has a frame in the shape of a rectangle. Several wires hold rows of movable beads. Young children may use them as toys.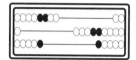

The idea came from the Russian counting board. A counting board is called an *abacus*. It is a type of calculator, invented long before the calculators we use today. The abacus above shows the number 815.

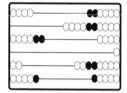 Here is a simple Russian abacus. It shows the number 406.25. Count the beads on the left in each row. How many beads are on each wire? What do you notice about the colors of the beads? Where is the decimal point?

1. Read the number shown on each drawing of a Russian abacus.

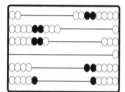 _____

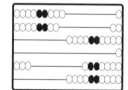 _____

2. Show each number on the diagram of the abacus.

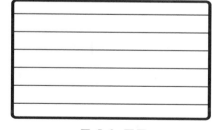

624.50   **541.75**

Now Try This

☞ Pretend that you are visiting Russia. You notice in the stores that an abacus lies next to the cash register. Write a letter to a friend back home describing the abacus. You may want to draw a picture.

THE CHINESE ABACUS

The Chinese call their abacus a *suan pan*, meaning "counting board." You may have seen such an abacus in Chinese shops or restaurants. It lies flat on a counter or table. This abacus shows the number 60,347.

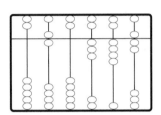

Can you see how it works? Look at the beads that have been moved to the crossbar. What do the beads above the bar stand for? What do those below the bar stand for? What does each string stand for?

1. Read the number shown on each drawing of a suan pan.

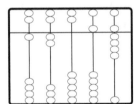

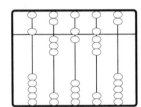

_____   _____

2. Show each number on the diagram of the Chinese abacus.

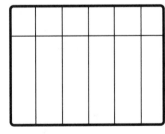

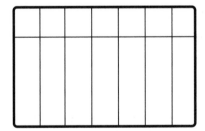

56,830   407,921

Now Try This

☞ The suan pan is like using the fingers of one or two hands—grouping by fives and tens. How does it work? How does it relate to Chinese stick numerals? Write your answers on the back of the sheet.

Part II

USING NUMBERS IN REAL LIFE

The seven lessons in this section describe the use of numbers in many societies, ancient and modern, including our own.

Lesson 7 is about money. When did the need for money arise? Some early forms of money, such as beads and shells, served other purposes as well. Students carry out transactions with these currencies, and are encouraged to calculate mentally.

Lesson 8 is about the *quipu* of the Incas, a set of knotted strings that might have told a whole history. Students reinforce their understanding of place value as they keep records on their own quipus.

Lesson 9 features the Roman linear measures that inspired the customary system still in use in the United States (but in few other countries), and the Egyptian cubit, perhaps the very first standardized measure. Lesson 10 deals with the relation between a linear dimension and the area of a pizza.

Lesson 11 is about the thirteen-month calendar of the Iroquois Indians and the three calendars that the Maya used simultaneously. Students review Maya numerals and learn to use them in a new way.

Lesson 12 departs from the "useful numbers" category to dip into the subject of magic squares in Chinese and Arabic cultures. Students learn to decipher East Arabic numerals and to construct their own magic squares.

Lesson 13 returns to the present and our own endangered environment. Students perform calculations to deal with such questions as saving water and energy and recycling garbage.

Book Links for Students

Ashabranner, Melissa, & Brent. *Counting America: The Study of the United States Census.* New York: Putnam, 1989.

Bruchac, Joseph & Jonathan London. *Thirteen Moons on Turtle's Back.* New York: Philomel, 1992.

Burns, Marilyn. *This Book Is About Time.* Boston: Little Brown, 1978.

Cribb, Joe. *Money* (Eyewitness Books). New York: Knopf, 1990.

Earthworks Group. *Fifty Simple Things Kids Can Do To Save the Earth.* Kansas City: Andrews and McMeel, 1990.

Elkin, Benjamin. *Money.* Chicago: Children's Press, 1983.

McMillan, Bruce. *Eating Fractions.* New York: Scholastic, 1991.

Myller, Rolf. *How Big Is A Foot?* New York: Dell, 1991.

UNICEF. *The Little Cooks.* Order from U.S. Committee for UNICEF: (800)553-1200.

Watson, Clyde. *Tom Fox and the Apple Pie.* New York: Crowell, 1972.

MONEY

Getting Started

Ask students to think about different times when they used money. You might begin by discussing that today we think of money as paper bills and metal coins, not to mention checks and credit cards. In the past, *barter* (exchange of goods) satisfied the needs of society until trade expanded to the point where money became necessary. Introduce the term barter.

Extension Activities

1. Students might discuss and dramatize the types of exchanges between Native Americans and colonists that would involve money. How did the two societies regard money differently?

OBJECTIVES FOR STUDENTS

♦ To learn how the need for money developed.

♦ To become familiar with some of the great variety of objects that have been used as money in various societies: beads, shells, silver, paper.

♦ To carry out calculations with some of these currency objects.

2. Students can act out a market scene in West Africa in the 19th century and pretend to use cowrie shells as currency.

3. Assign different groups of students to research the currency of various societies and report to the class. They might make posters to illustrate their research.

4. Students might look up the currencies of several countries in the daily newspaper. What are their values as compared to the U.S. dollar? Does the relationship change over time? What would be the cost of a certain item in Japan or Mexico or Canada?

5. When the Spanish conquerors reached Mexico in 1519, they found that the Aztecs were using cocoa beans as currency. The largest unit was a bag of 8,000 beans. Ask students to relate this number to the Aztec base-twenty numeration system. How might this bag compare in weight to the West African bag of 20,000 cowries?

Answers to Student Pages

Page 39:
a) 24 + 12 = 36   b) 100 + 216 = 316
Now Try This) A quarter; 12; 26.

Page 40:
1) 75; 113.
2) 6 matanu; 10 matanu, 4 strings, 3 beads.
3)

Units	Number of Shells
String	40
Bunch	200
Head	2,000
Bag	20,000

Now Try This)   a. 1,725   b. 42,160.

OBJECTS FOR MONEY

Early forms of currency usually had intrinsic value—as useful objects, as decoration, for ritual purposes. Some of these early forms were beads, cowrie shells, textiles, and metal objects. The first coins were made in Lydia (now part of Turkey) about 26 centuries ago.

Name: _____

🎧 WAMPUM AND PIECES OF EIGHT

A long time ago, before people used paper currency, they used a system of exchange (trading) known as **barter**. But barter is not always convenient. That's why people began to use **money**.

Objects that had value in themselves became the first type of money used. For example, the Iroquois (Indians) of the Northeast made shell beads, called **wampum**. A design made of wampum beads might have a certain message. A wampum belt might record a treaty between nations. Later the Iroquois sold fur pelts to European colonists for strings of wampum beads. The colonists also used wampum as their money.

Suppose a beaver pelt costs four strings of wampum beads and an otter pelt costs three strings. Find the cost of:

a. Six beaver and four otter pelts.
b. 25 beaver and 72 otter pelts.
c. Make up other problems about wampum.
    Ask your classmates to solve them.

Now Try This ━━━━━━━━━━━━━━━━━━━━

☞ A long time ago Spain had a popular silver coin worth eight *reals*. People called them Spanish dollars or "pieces of eight." If they needed a smaller coin, they would cut up a dollar. A fourth of the coin was called "two bits." Later they used the same name for one-fourth of a U.S. dollar.

What U.S. coin is worth "two bits"?
How many **bits** are equal to $1.50? $3.25?

Name: _____

ⓐ BEADS AND SHELLS IN AFRICA

In the Congo region of Africa, beads were used as currency. Five beads made a "string." A group of five strings was called a *matanu*.

1. On a separate sheet, calculate the number of beads (show your work):

a. 3 matanu

b. 4 matanu, 2 strings, 3 beads

2. How many matanu, strings, and single beads would you pay for an object that cost:

a. 150 beads? _____

b. 273 beads? _____

In parts of West Africa, a century or more ago, 40 cowrie shells were strung on a necklace called a "string." Five strings formed a "bunch." Ten bunches were called a "head." Ten heads made a bag of 20,000 shells.

3. Complete the table below using the information given above.

Units	Number of Shells
String	
Bunch	
Head	
Bag	20,000

Now Try This

☞ In West Africa, women with little or no schooling can compute the largest of sums in their heads—mentally. Compute (mentally, if you can) the number of cowrie shells in:

a. 8 bunches, 3 strings, 5 shells

b. 2 bags, 1 head, 4 strings

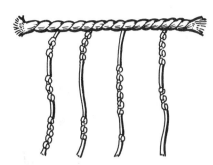

KEEPING RECORDS

Background

The Inca Empire flourished from about 1400 to 1560. At its peak it encompassed a vast territory along the western part of South America. Cuzco, its capital city, was situated at an elevation of 11,000 feet in the land that is now Peru. The common language was Quechua, still spoken today, although the people of the subject lands spoke many other languages.

The Inca empire was well organized; the rulers had information at their fingertips. This information was encoded in *quipus*. A quipu is a collection of colored strings in which knots have been placed to indicate various quantities, using a base-ten place-value number system. The knots in the units place were formed differently from those in the higher-valued positions. The color and placement of the strings furnished the key to the categories of statistics. Quipu makers were trained at special schools in Cuzco.

> ### OBJECTIVES FOR STUDENTS
>
> ♦ To learn how the Inca kept records with knotted strings.
>
> ♦ To do addition with knots in strings.
>
> ♦ To make a quipu.

Getting Started

Begin by discussing the kinds of numbers that are important to people and the different ways people remember them. Students might mention age and birthdays, phone numbers, addresses, Social Security numbers, license numbers, etc. which are necessary for everyone. Merchants need to know prices and quantities. The government collects statistics of all kinds—census data, etc. Records may be kept on paper or in computers.

For planning and making the quipu each student should have colored pencils and two different-colored strands of heavy cord or yarn, about 50 cm (20 inches)

long, as well as a shorter cord to tie them together.

Ask students to locate the Inca Empire on the map. They might relate the Inca era in history to the coming of the Spanish conquerors.

Extension Activities

1. List the ways that the quipu system is like our pencil and paper numbers, and the ways it is different. (See box for answer.)

2. The Inca government sent runners (messengers) to collect information on quipus and bring them to Cuzco, the capital of the Inca Empire. Work with your group to write a skit about this. What do you think was done with the quipus in Cuzco?

3. Students may want to research the use of knots as numerals in other cultures.

4. How are data collected and stored in our culture? Students might read Ashabranner (1989) on the U.S. census.

> ### STRINGS THAT COUNT
> Quipu numbers are like Indo-Arabic numerals in that both use base-10 positional notation. They are different in that they are in three dimensions rather than on a flat surface; knots are counted instead of having a different symbol for each digit from zero to nine; zero is shown by an empty space.

Answers to Student Pages

Page 43:
1) 235; 403; 61; 330.
2)

Page 44:
1)

Name: _____

[9] KNOTS ON A STRING

How do you remember important numbers, like your birthday? You probably memorize them. But suppose you have to remember many numbers? The Inca people of South America had a special way of remembering numbers. They made knots in strings of different colors. A set of knotted strings tied to a main cord was called a *quipu*.

Inca numbers were grouped by tens, like the Indo-Arabic numerals we use. The diagram shows four cords tied to a main cord. The highest value is closest to the main cord. Look at the cord on the left. It has three groups of knots:

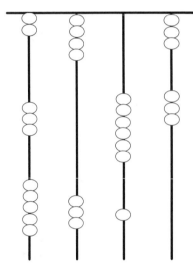

- the top part of the cord is the hundreds place;
- the middle part is the tens place;
- the bottom part is the ones place;
- a space with no knots stands for zero.

1. Read the numbers on each cord, starting from the left

2. On a separate sheet, draw a quipu with four cords hanging from the top cord. Show these numbers on the cords, starting from the left: 172; 301; 56; 230.

Now Try This

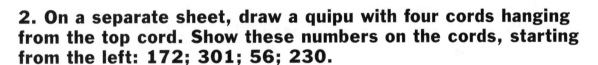

☞ Make a plan of a quipu having two cords. One cord shows 23 and the other shows 14. Then use cords of two different colors and make a quipu like the one you drew.

Name: _____

⒜ MORE KNOTS ON A STRING

The Inca government used quipus to keep account of people, livestock, and products, to keep records of taxes, to recall events, and for many other purposes.

This diagram shows a quipu with a top cord attached to the main cord. The top cord gives the sum of all the numbers on the lower cords. It shows the places "upside down." The highest place is nearest to the main cord.

1. Draw the correct knots on the top cord. Try to do the addition without using paper and pencil. How can you show a number in the thousands place?

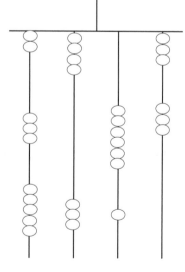

An Inca village needs to know how many potatoes each family will use in a month.

2. On a separate sheet, draw a quipu having three cords of different colors. Each cord is for a different family. "Knot" the cords to show how many potatoes each family ate. First make a plan. Exchange quipus with a classmate and read each other's quipu. Write down the number shown on each cord.

Now Try This

☞ Add a top cord to your quipu, showing the total number of potatoes for the three families. Try to do the addition without using paper and pencil.

How Big?
How Small?

Background

The ancient Egyptians may have been the first people to adopt standard units of measurement. The construction of the Great Pyramid of the Pharaoh Khufu about 4,600 years ago necessitated the use of standard units. The *cubit* (distance from the elbow to the fingertip) used in construction was named the royal cubit. A shorter cubit was used in other contexts. The *hayt* was 100 cubits. The *double remen* was the length of the diagonal of a square whose side measured one royal cubit.

> ### OBJECTIVES FOR STUDENTS
>
> ◆ To understand that measurement generally originated with reference to the human body.
>
> ◆ To understand the need for standard units of measurement.
>
> ◆ To learn the units of linear measurement of the ancient Romans and Egyptians.

Later societies continued to use the cubit, exemplified by the Bible. The Greek cubit was about 18 inches and the Roman, 17.5 inches.

Like the Egyptians, the ancient Romans were great builders and engineers. Some of their buildings still stand today, two thousand years later. It is believed that the Romans developed the unit of measure called a foot. In their language, called Latin, the word for foot is *pes*. For two or more feet they said *pedes*. The width of the thumb was *uncia*. (The English word "inch" comes from *uncia*.) Twelve of these units equaled a *pes*.

Everyone had to agree on the length of the units called foot and inch. They needed *standard* measures.

Getting Started

As preparation for this lesson, students might discuss how people use parts of their body to show size or distance (one way to approximate a yard is by extending the material along your extended arm to the tip of your nose). Some students may ask family members or other adults if they use this method, or any other. Students can develop a "feel" for standard units by comparing them with various parts of the body.

Students might suggest extending their hands and holding them a certain distance apart. Ask whether this measure is exact. Students should understand that no measure is exact, but some measures are more precise than others.

Extension Activities

1. Can you think of any English words that sound like the Latin words for foot or feet? Write the words and their meaning.

2. Students might write skits showing the disastrous effects of using nonstandard units of measurement in situations calling for standardization. See Myller (1991) in the bibliography on page 36.

3. Students might research the development of units of measure. Many of the references, as well as the encyclopedia and adult history of mathematics books, include this subject. Various aspects of the topic may be assigned to different groups of students.

Answers to Student Pages

Page 48:
1) Answers will vary. The ratio of the royal cubit to the personal cubit should be more than one and less than two, unless the student is unusually tall or short.

2) A palm = 7.4 cm.

Now Try This) Answers will vary, depending upon the objects measured.

Name: _____

Ⓐ ROMAN MEASURES

Have you ever used your hands to show how big something is? Have you used any other part of your body to measure size or distance?

Long before rulers and tape measures were invented, people measured things by using the parts of their body. We still use the word "foot" for a certain unit of measure. Is your foot as long as the unit we call a foot?

Work with a partner. Use your own foot as the unit of measure. Walk a distance of ten feet, heel-to-toe. Mark the starting and finishing points. Then measure the distance with a foot ruler.

How many standard feet did you walk?

Ten of my feet = _____ standard feet.

Now Try This

☞ Work with a partner. Make a ruler that is as long as your foot. Ask your partner to measure your height using your own foot ruler. Then do the same for your partner, using his or her "foot" ruler. Compare your numbers.

My height is _____ times the length of my foot.

Name: _____

ⓐ THE EGYPTIAN CUBIT

The ancient Egyptians built impressive temples and other buildings. For this, they needed accurate measurements. Many of their measurements were based on parts of the body, like the "foot," a word that is still in use. Other measures were the palm, handspan, and **cubit**. The cubit was the distance from the elbow to the fingertip. To build the Great Pyramid, architects used the royal cubit, 52 cm. (21 in.) long, divided into seven palms or 28 digits. It was measured out on a slab of granite and used as a standard for the cubit stick.

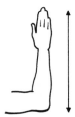

1. Measure your own cubit and compare it with the royal cubit.

My personal cubit measures _____ cm., to the nearest cm.

The royal cubit is about_____ times as long as my personal cubit.

2. Work with a partner. Make a cubit stick equal in length to the royal cubit. Divide it into seven palms.

Each palm is _____ cm. long (nearest tenth).

Now Try This ━━━━━━━━━━━━━━━

☞ Place a hand on a piece of paper so that the pinky side lines up with the paper edge. Mark a pencil dot on the opposite side. Fold the paper at the dot. The edge of the paper is your palm measure. Estimate the length of several objects, such as your desk, chair, and notebooks, in palms. Write your estimate. With your partner, measure each object. Write your measurement in palms. Complete the table.

Object	Dimension	Estimate (palms)	Measurement (palms)

COOKING

Background

We commonly associate tomatoes with traditional Italian foods. Yet this common fruit was unknown outside of America until Europeans found that the indigenous peoples of the lands that are now Peru and Mexico, were raising tomatoes that varied in color and ranged from cherry to almost melon size. For years Europeans regarded tomatoes as poisonous. Eventually they formed part of many new delicious dishes. Other common food crops such as potatoes, corn, peanuts, and various kinds of squash also originated in the Americas. Students should locate Italy, Peru, and Mexico on the map.

Getting Started

In preparation for this lesson, students might discuss dishes that incorporate tomatoes, and ask their families to supply additional information. Perhaps some students are not familiar with pizza, the Italian word for pie.

Extension Activities

1. Students can work in groups to try to find out where pizza ingredients for the crust and toppings came from. Tomatoes, for example, were first grown by the native peoples (Indians) of America.

OBJECTIVES FOR STUDENTS

♦ To relate areas of plane figures to their linear dimensions.

♦ To discover that a square is a convenient unit to measure area, while a circle is not.

♦ To change a recipe to accommodate various numbers of consumers.

♦ To learn the origin of some common foods, such as tomatoes.

49

2. Students can use pizza to learn fractions.

3. Students might look at the health value of pizza ingredients and other foods, and present their findings in a mathematical format. A good source of information is KAJF (Kids Against Junk Food), c/o CSPI (Center for Science in the Public Interest), 1875 Connecticut Avenue, Washington, D.C. 20009; (202)332-9110.

> **PIZZA, PIZZA**
> Pizza is the Italian word for "pie." The term "pizza pie" uses the same word twice and is redundant.

4. Write a story and draw pictures to show the many uses of tomatoes. Don't forget pizza. With your group, make a poster to show the class.

Answers to Student Pages

Page 51:
1) More than twice the area.
2) Four small squares fit into the large square; four.
3) Four. A circle is not a good unit to measure area because circles laid end to end leave empty spaces, squares laid end to end leave no empty spaces and do not overlap. Now Try This) Four times two ounces, or eight ounces.

Page 52:
1)
Servings	Water (cups)	Yeast (oz)	Flour (cups)	Oil (tsp)	Salt (tsp)
8	2	1/2	5	2	2
12	3	3/4	7 1/2	3	3
14	4	1	10	4	4
16	4	1	10	4	4

2) To compare the eight-inch and twelve-inch pans, compare 8 x 8 (64) with 12 x 12 (144), which reduces to 4/9. Thus an 8-inch pan requires slightly less than half the ingredients, but since all the given measurements in the recipe are only approximate numbers (that point should be emphasized), half the recipe should produce excellent results.

Name: _____

🖉 MAKING PIZZA

Ed and Maria are making pizzas for a class party. They know that *pizza* is the Italian word for pie. Maria makes small pizzas with different fillings. Ed makes larger pizzas, twice as big across (diameter) as Maria's.

1. If Maria's pizza is enough for one portion, how many portions are there in Ed's pizza? Ed says two, but Maria thinks there are more. What is your guess? (Don't do any work on paper.)

Ed and Maria think the pizzas would be easier to compare if they were square. Maria cuts a four-inch (10 cm) square and Ed cuts an eight-inch (20 cm) square out of paper. Cut squares like these and compare them.

2. How many small squares fit into the large square? _____. How many small squares make a square that has a side twice as long? _____.

3. Use the paper squares in activity 2. Cut a four-inch circle from the small square and an eight-inch circle from the large square. Fold your square into quarters before you cut. How many small circles fit into the large circle? _____. Why is it harder to compare circles than to compare squares?

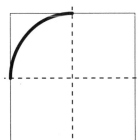

Now Try This ━━━━━━━━━━━━━━━━

☞ The children decided to use tomato paste and other fillings. A four-inch pizza needs two ounces of tomato paste. How much is needed for an 8-inch pizza? Write your explanation.

Name: _____

MAKING MORE PIZZA

Kimara and Jon are planning a pizza party. They
have a recipe for a pizza crust that will serve four people. It requires a pan with a diameter of 12 inches
(about 30 cm). They are not sure how many people
will attend the party. They want to calculate the
quantity of ingredients and the number of 12-inch
pans they will need for various numbers of people.
They plan to use several varieties of toppings.

These are the ingredients for a pizza crust that will serve four people:

1 pkg (1/4 oz) active dry yeast   1 cup warm water
1 tsp olive oil for the pan   1 tsp salt
2 1/2 cups all-purpose wheat flour

**1. Calculate the quantities of each ingredient for the following
numbers of people: 8, 12, 14, 16. Remember, you only have 12-
inch pans, which serve four people. Arrange your answers in the
form of a table (chart).**

**2. Now suppose you have only 8-inch pans. Calculate about how
much of each ingredient you will need for one crust. What
method will you use to make the comparison between the two
sizes of pan?**

Now Try This

☞ One topping will consist of tomatoes and cheese, topped with slices
of pepperoni (sausage) and olives. Draw a picture of this pizza, with
the toppings arranged so that it can be divided into four equal slices
and each slice will have the same amount of toppings.

CALENDARS

Getting Started

Begin this section by carefully looking at a calendar with your students. Challenge students to discuss everything that a calendar tells them.

Turn the discussion to the number of days in the cycles of the moon and the sun. The moon has a cycle of about 29 1/2 days, counting from one new moon to the next. Twelve lunar cycles add up to 354 days. The sun has a cycle of about 365 1/4 days. The quarter day is made up every four years (with certain exceptions) by an extra day in a leap year. Throughout history peoples have devised different ways to reconcile the discrepancy between the lunar and solar cycles.

OBJECTIVES FOR STUDENTS

◆ To understand that the year can be subdivided in various ways.

◆ To work with lunar and solar calendars.

◆ To compare our official calendar with other calendars.

Extension Activities

1. Ask students to research the calendars of other cultures, e.g., Chinese and Egyptian.

2. Find more information about how dates are set for various special days, such as birthdays of famous men and women, Election Day, etc.

3. Read about the Ishango bone, recently redated to about 25,000 years ago, and thought to represent a six-month lunar calendar (Zaslavsky, 1979).

4. Student can write down the ways that the Iroquois calendar is like ours and how it is different.

Answers to Student Pages

Page 55:
1) 364 days.
2) 365, except for 366 every four years.
3) January, April or May, and September.

Page 56:
1) 365 days. Our year is the same, except for leap years.
2) 10 years, 7 months, 19 days = 3,600 + 140 + 19 = 3,759 days.
Now Try This) Answers will vary. Describe the three types of Maya calendars discussed in the box on this page.

> ### THE MAYA CALENDAR
> The Maya "year" had 365 days for everyday use, 260 days for ritual or sacred purposes, and 360 days for the recorded passage of time. The possible reasons for the 20-day periods of time in the Maya calendar are linked to their base-20 number system.

Name: _____

⌒ THE IROQUOIS CALENDAR

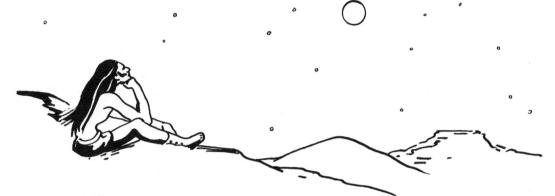

The Iroquois live in the northern part of New York State and in some parts of Canada. They call themselves the Haudenosaunee. "Iroquois" is the name given them by the French. Long ago they worked out a calendar by watching both the moon and the sun. Their year had 13 moons. Each moon had 28 days.

1. How many days were in the Iroquois year? _____ days.

2. How many days are in our year? _____ days. Write what you know about the months in the calendar we use.

3. Each Iroquois moon had a name. The first moon was called Nis-ko-wok-neh, the moon of snow and blizzards. The fifth moon was Wen-taa-kwo, the flowers. The ninth moon was Ke-to-ok-neh, the harvest. What months in our calendar might have those names?

Now Try This ━━━━━━━━━━━━━━━━━

☞ Make a list of our months and give each a name according to the weather or an important activity. Draw pictures to show what happens in each month.

Name: _____

⊙ THE MAYA CALENDAR

The Maya live in the region that is now southern Mexico and northern Central America. The Maya and their neighbors had more than one calendar. For everyday purposes they used a calendar of 18 periods that we will call months. Each month had 20 days. Five days were added at the end.

1. Using the information given above, how many days were in the Maya year? _____ **How does it compare with our year?**

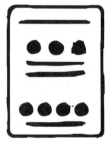

When the Maya inscribed periods of time on their stone monuments, they used a different system. For this purpose the year had 360 days—18 months of 20 days each. Read the number to your left as: 5 years, 13 months, 9 days to calculate the number of days: $(5 \times 360) + (13 \times 20) + (9 \times 1) = 2,069$ days.

2. Using the monument to your right, show how you calculate the number of days for the period of time.

_____ years, _____months, _____days

= _____ + _____ + _____ + _____ days

Now Try This ━━━━━━━━━━━━━━━━━━

☞ Work with a partner. Write two Maya calendar numbers for your partner to calculate. Is your partner correct? Research the three types of Maya calendars. How do we set the dates for holidays like Labor Day or New Year's Day? Discuss the dates for holidays in different religions and cultures.

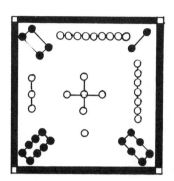

MAGIC SQUARES

Getting Started

Magic squares offer an excellent opportunity to practice mental arithmetic, as well as to discover patterns. Students can carry out many activities in addition to those described in the lessons, as suggested in these notes.

You may wish to work through the first part of the Ancient China reproducible on page 59 with the class, to make sure everyone understands the principles of magic squares.

Extension Activities

1. Students can research other games popular with these cultures.

2. Challenge students to make their own magic squares.

3. Ask students to pretend that they live in Ancient China. They have just witnessed the turtle with the magic pattern emerge from the water (story on page 59). How would they describe in a letter to a friend what they just saw?

4. Students can research other Chinese legends.

FEATURES OF THE CHINESE MAGIC SQUARE

The middle number is 4. The consecutive numbers 3, 4, and 5 constitute one diagonal. Odd numbers are in the corners, even numbers in the center and sides. Students may note how the squares are related by symmetry. This property becomes clear if a large square is drawn on translucent paper and the numbers represented as dots, as in the Chinese magic square. By rotating and flipping this square, students can obtain other magic squares. There are eight different squares.

Answers to Student Pages

Page 59:

1)
4	9	2
3	5	7
8	1	6

2) The magic sum is 15.

3)
5	0	7
6	4	2
1	8	3

3	2	7
8	4	0
1	6	5

5	6	1
0	4	8
7	2	3

Page 60:

1)
8	3	4
1	5	9
6	7	2

2)

1	2	3	4	5	6	7	8	9

3)
4	9	2
3	5	7
8	1	6

8	1	6
3	5	7
4	9	2

Name: _____

ⓐ ANCIENT CHINA

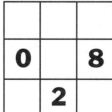

The Chinese tell this story. About 4,000 years ago the emperor and his court were sailing down the River Lo. Suddenly a turtle appeared out of the water. On the turtle's back was the design you see here. When they looked at it carefully, they saw that nine numbers were in the design.

1. Using the pattern on the turtle, count each set of black dots and place the correct numbers in the corners of the empty square on the left. Then, count each set of white dots and place the correct numbers in the center and sides of the empty square.

2. Add each row across. The sums are: first row ____; second row ____; third row _____. Add each column going down. The sums are: first column _____; second _____; third _____. Add each diagonal. The sums are ____ and ____.

The sum is always the same number. This number is called the *magic sum*. The square is called a *magic square*.

3. Complete the magic squares below so that the magic sum is 12. Use each of the numbers from 0 to 8 exactly once. Check all eight sums in each square

5	0	
	4	
	8	

3		7
1		

0		8
	2	

Now Try This ━━━━━━━━━━━━━━━━

☞ Make your own magic square using the numbers 2 to 10 once each.

Name: _____

a EAST ARABIC NUMERALS

Muslim scholars enjoyed making large, complicated magic squares as shown below. They called this the "science of secrets." They wrote in East Arabic numerals.

The square to the right uses the numbers from 1 to 9 exactly once each. The magic sum is 15—the sum of every row, every column, and each diagonal.

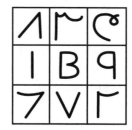 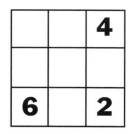

1. Translate this square into our Indo-Arabic numerals. There are some hints to help you. Check that you have the magic sum 15 eight different ways.

2. Write the East Arabic numerals that match the Indo-Arabic numerals we use:

East Arabic numerals		┌		℃		7		
Indo-Arabic numerals	1	2	3	4	5	6	7	8	9

3. Use your chart in #2 to translate these magic squares:

 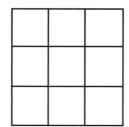 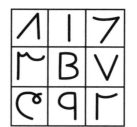 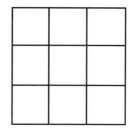

SAVE THE EARTH

Getting Started

Introduce this lesson on conservation by reading and discussing one or more stories from Caduto & Bruchac's *Keepers of the Earth* (Fulcrum, 1988) or a similar collection to illustrate the Native peoples' regard for the earth and its resources.

In the afterword to *Keepers of the Earth*, Joseph Bruchac writes about the Native American peoples' "continent-wide belief that mankind depended on the natural world for survival, on the one hand, and had to respect it and remain in right relationship with it, on the other."

Calculators are recommended for most of the exercises.

OBJECTIVES FOR STUDENTS

◆ To recognize the waste of resources in our society.

◆ To discover Native peoples' contrasting attitudes to the environment.

◆ To solve complex mathematical problems dealing with conservation.

◆ To learn why they should take action to save our resources.

Extension Activities

1. Ask students to work in groups to find out how much water they can save by turning off the tap when they brush their teeth, wash dishes, etc. Each group can make a poster showing how they can save water. Some may choose to write a story about their research.

2. Students can choose one kind of material that people throw away and do a little research. Questions to consider: About how much of this material did you and your family throw away in the past week, at work and at school? How could you have saved or recycled some of this material?

Answers to Student Pages

Page 63:

1) Answers will vary.

2) 2 x 60 sec = 120 sec; 120 ÷ 3 = 40; 40 cups (2 1/2 gallons).

3) 39 cups; Answers will vary.

Now Try This) 546 cups (34 1/8 gallons) in a week; 28,392 cups (1,774 1/2 gallons) in a year. The figures should be rounded to indicate that the original measurements were only approximate: about 34 gallons in a week and 1,800 gallons in a year.

Page 64:

1) a. 260 cans.

b. Each set would run for 975 hours, or 40 days and 15 hours if the sets ran continuously.

c. Approximately three billion gallons.

HOW ACCURATE IS IT?

No measurement is exact, no matter how precise the tools used to measure.

Name: _____

[◖] SAVE THAT WATER!

Does anyone in your family let the tap water run while she or he is not using it? Do you turn off the tap while you brush your teeth?

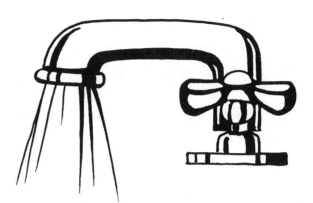

1. Guess how much water you save when you turn off the tap every time you brush your teeth, instead of letting the water run. Write your answer on a separate sheet.

Zena and Chen decided to find out. They opened the tap and counted how many seconds it took to fill an 8-ounce cup with water—three seconds. Then they timed how long it took Zena to brush her teeth—two minutes.

2. How much water was running while Zena brushed her teeth? Show how you did the work. You may use a calculator.

3. Zena really needed only one cup of water to brush her teeth. How much water was wasted by leaving the tap open? Compare this amount with your estimate in activity 1.

Now Try This ━━━━━━━━━━━━━━━━━━━

☞ Zena brushes her teeth twice a day. How much water does she waste by leaving the tap open:

In one week? _____

In a year? _____

Name: _____

ⓐ SAVE THOSE CANS!

Let's face it. People waste a lot of stuff—paper, food, cans. If we're not careful, the world may run short of these things one day.

Aluminum soda cans should be recycled. You may not realize how much energy goes into making one soda can. Here are some figures:

* The United States uses 65 billion aluminum soda cans per year.
* Throwing away one can wastes six ounces of gasoline.
* Energy saved from recycling one can will keep a television set running for three hours.

1. Using the information given and a calculator, answer the questions on a separate sheet.

a. There are about 250 million people in the United States. How many soda cans does each person use, on the average?
b. Assume there are 200 million TV sets in the U.S. How many days would they run on the energy saved by recycling all soda cans in one year?
c. There are 128 ounces in a gallon. How many gallons of gasoline would be saved per year if all cans were recycled?

Now Try This ━━━━━━━━━━━━━━━━━━

☞ Discuss with your group a project for recycling soda cans. You might gather statistics on the number of cans your group and their families use, or are sold in school or in the local supermarket. How many are recycled? Make graphs and posters to illustrate your work.

Part III

SPACE,
SHAPE,
AND SIZE

The six lessons in this section deal with art and architecture in several cultures. Students will apply their knowledge of geometry, measurement, and symmetry as they create their own works of art.

In lessons 14 and 15 students consider the relationship between the area and perimeter of several types of plane figures—square, rectangle, circle, triangle. The contexts include the traditional American log cabin, the longhouse of the Iroquois (they call themselves the Haudenosaunee, "people of the longhouse"), and a round house in Kenya. Students use their imaginations to create and decorate models of these homes.

Quilting is the theme of Lesson 16. Children become familiar with quilting patterns and then create their own, first on paper and then in cloth. They learn that quilting was an art form practiced by American women of both African and European origin. Families and community members can share their expertise and overcome their fear of math as they work with the children.

Lesson 17 introduces the *adinkra* cloth of Ghana. Originally worn at funerals, adinkra features a variety of motifs, each with its own significance, repeated within rectangular borders. Students analyze the designs for both line symmetry and rotational symmetry, and create their own adinkra-like patterns.

Repeated geometric patterns are characteristic of the art associated with the Islamic faith. In Lesson 18 students create tessellations with pattern blocks and on paper, and determine which geometric shapes cover a surface without overlap or empty spaces. Angle measurement is a key.

Through Native American beadwork in Lesson 19, students analyze the relationship between the widths (or any linear dimensions) and the areas of similar shapes. They hone their calculating skills as they compute the number of beads required for patterns of different sizes but the same shape.

Book Links for Students

Corwin, Judith Hoffman. *African Crafts*. New York: Franklin Watts, 1990.

Flournoy, Valerie. *The Patchwork Quilt*. New York: Dutton, 1985.

Grifalconi, Ann. *The Village of Round and Square Houses*. Boston: Little, Brown, 1986.

Hunt, W. Ben. *Indian Crafts and Lore*. Golden Press, 1954.

Ridington, Robin & Jill. *People of the Longhouse: How the Iroquois Lived*. Buffalo, NY: Firefly, 1992.

Yue, Charlotte & David. *The Tipi*. Boston: Houghton Mifflin, 1984.

ARCHITECTURE

Getting Started

You might introduce this lesson by asking students to describe the shapes of the buildings they know, particularly their own homes. Have them concentrate on the floor plan and the rooms into which the home is subdivided. Do the walls follow straight lines and do they meet at right angles? Are students familiar with exceptions to this conventional style?

For The Log Cabin, students (or groups) will each need 16 square manipulatives. For The Iroquois Longhouse, students will need grid paper and tape measure or measuring rods.

OBJECTIVES FOR STUDENTS

♦ To perform calculations involving the perimeter and the area of a rectangle.

♦ To discover that the rectangle with the greatest area for a given perimeter is a square.

♦ To learn about the construction of log cabins and the Iroquois longhouse.

♦ To consider the factors that influence the style of a home.

Extension Activities

1. Discuss with students the factors people have to consider when they plan to build their homes.

2. Conduct a survey of the community to classify houses and record the number of each. Students should first agree on the categories of buildings and the territory to be surveyed. Then they might graph the results and write up their conclusions.

> **TEACHER TIP**
> The area and perimeter of a 4 x 4 square has the same number of units. The only other rectangle with this property is a 3 x 6 rectangle.

3. Students can investigate how many different rectangles they can form with 25 square units.

4. Invite students to investigate homelessness in the community and discuss possible solutions to this problem.

5. Students can research rectangular houses in other societies.

Answers to Student Pages

Page 69:

1)

Rows	Squares in each	Distance around
1	16	34 units
2	8	20 units
4	4	16 units

If students know the length of the edge of the square manipulative (e.g. 1 cm or 1 in), they might use this unit instead of the word "unit."

2) 4 x 4; square; 16 square units; 16 units. Now Try This) A cabin built on a square foundation has the largest area for a given amount of material for the walls. Students might consider exposure to the sun, winds, rain, and snow; the number, shape, and location of the rooms; method of heating, etc.

Page 70:

1)

Model	Width	Area	Perimeter
A		750 sq ft	130 ft
B		1,080 sq ft	156 ft
C	19 ft	1,425 sq ft	
D	22 ft		244 ft

Name: _____

 # THE LOG CABIN

It is the year 1800. Some families have just come to Kentucky from the state of Virginia. All around them is forest. They must cut down trees to build log cabins. They must do all the work themselves, and it is not easy. How can we help them?

They decide to build the house so that the floor is in the shape of a rectangle. They want to make the biggest rectangle they can. They don't want to cut down too many trees. Find the shape of the best rectangle.

1. Use 16 squares of the same size. Arrange the 16 squares in the shape of a rectangle in as many ways as possible and fill out the chart below. The first one is done for you.

Rows	Squares in each	Distance around
1	16	34 units

2. Find the distance for each rectangle.
Which rectangle has the smallest distance around?
What is the special name for this rectangle?
What is the area of this rectangle?
What is the perimeter of this rectangle?

Now Try This

☞ Cutting down trees is hard work. The families want to cut as few as possible. What shape should they use for the floors of their homes if they want the largest floor space? What else must they think about when they plan the shape of their homes?

Name: _____

THE IROQUOIS LONGHOUSE

The Iroquois live in northern New York State and in southern Ontario and Quebec. They call themselves the *Haudenosaunee*, which means "People of the Longhouse." Years ago several families lived together in a longhouse. Today the longhouse is used mainly for ceremonies.

The construction of a longhouse had to be planned carefully. It had a central hallway running the length of the house. A room for each family led off the hallway. The building might measure anywhere from 50 feet to 100 feet or more in length.

You have been assigned the task of building a longhouse for the village. You are designing several different floor plans. On grid paper, draw the floor plan for each model. Complete the table. Decide the number of rooms in the house and draw them on your floor plan. Label all the dimensions and the area.

MODEL	LENGTH	WIDTH	AREA	PERIMETER	WIDTH OF HALLWAY
A	50 ft.	15 ft.			3 ft.
B	60 ft.	18 ft.			4 ft.
C	75 ft.			188 ft.	5 ft.
D	100 ft.		2,200 sq. ft.		6 ft.

Now Try This

☞ Work with your group to build a model of a longhouse using one of the floor plans. Decide on a convenient scale.

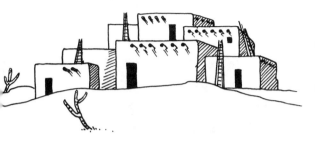

HOUSES

Getting Started

An excellent way to begin this unit is to pose the following questions to your students: Have you ever wondered why so many things in nature are round? Look at the moon, a tree, your finger. Look at the rings that people wear. Why are they round? What does the word "ring" mean?

Help your students make a list of things in nature that are round or nearly round. Then make a list of round things that people make. Discuss the possible reasons for making these things round. Some reasons you might offer: A tree or a finger, for example, grows out from the center so that every point on the circumference is roughly equidistant from the center. A ball offers less resistance to the atmosphere than a cube of the same weight. An eyeball or a wheel would not roll smoothly if it were any other shape. Some other human-made round objects are cans, dishes, buckets, lampshades.

OBJECTIVES FOR STUDENTS

♦ To carry out calculations involving linear dimensions and areas of plane figures.

♦ To discover that the circle is the largest shape having a certain perimeter.

♦ To use cylinders and cones in constructing models of round houses.

♦ To learn about societies that build round houses.

Extension Activities

1. Each group may choose one of the societies discussed on page 74 and research the factors in that society that influence the way they build their houses.

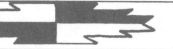

2. Ask students to imagine that their family is planning to build a house. They must gather the materials and do all the construction themselves. What shape should the materials be? Why?

Answers to Student Pages

Page 73:

1) Furniture is generally rectangular or nearly so, and would not fit well in a round room.

2) The true story of Rodah and her grandmother, with an accompanying photograph, is from Zaslavsky (1979), page 163. Students will enjoy reading Grifalcolni (1986), listed in the bibliography on page 66.

Page 74:

1) To accommodate rectangular furniture, layout of streets, etc.

2) Answers will vary. Ask students to justify their responses.

3 & 4) The circle should have the greatest area, about 80 sq. cm. The square is next, with an area of 64 sq. cm. The areas of the rectangle and triangle will vary according to the relative lengths of the sides, but will always be less than 64 sq. cm.

Now Try This) The tipi of the Great Plains in the U.S., the Central Asian yurt, many types in Africa, the Inuit igloo. Some factors that influence styles in housing are the environment, climate, lifestyle of the people, and traditions.

<table><tr><td>

THE REASON FOR ROUND TOWERS

Castles, which were generally rectangular, had round towers, so that attackers could easily be seen from any direction.
</td></tr></table>

Name: _____

THE ROUND HOUSE IN KENYA

Round houses are very popular in Kenya.

1. Imagine that a room in your house is round. Talk about the furniture and other things in that room. Would they fit as well in a round room? How would you feel in a round room?

Rodah lives in Kenya, a country in East Africa. Her family has just built a new square house. They invite Rodah's grandmother to stay with them for a while. She says: "You have a house with corners. I would get lost in such a house."

2. Talk about Rodah's grandmother's feelings. Can you understand them? Make up a skit about Rodah's family, their new house, and the invitation to her grandmother.

Now Try This

☞ Suppose you wanted to make a paper model of this round house with a conical roof. How would you make the round wall? How would you make the roof?

Name: _____

 # ROUND HOUSES

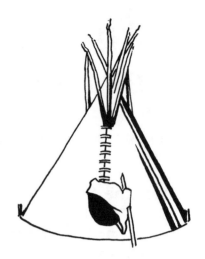

The tipi on the left is round. Can you think of a reason why? How about castles? Can you think of why they had round towers? Now, think of your school building, your classroom, your home. What is the shape of the building and its rooms? Most likely they are rectangular.

1. List the reasons that buildings are usually rectangular rather than round.

2. Suppose you have a choice of shapes for the base of your house—square, rectangle, circle, or triangle. You have only a certain amount of material for the walls. Guess which shape will give the greatest amount of floor space, and give the reasons for your guess.

3. Using a sheet of centimeter-squared paper and a piece of string 32 cm in length, draw the four shapes listed in activity 2. The perimeter of each shape is 32 cm. Find the area of each shape by counting the small squares. Write your results in the table.

Shape	Perimeter	Area
Circle	32 cm.	sq. cm.
Square	32 cm.	sq. cm.
Rectangle	32 cm.	sq. cm.
Triangle	32 cm.	sq. cm.

4. The shape with the largest area is _____.

Now Try This

☞ Name some societies that construct round houses. What factors influence the style of housing? Make a list and discuss them.

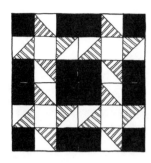

PATCHWORK QUILTS

Getting Started

Start by explaining that a long time ago women made the clothing for their families. And in many cases they saved left-over pieces of fabric to make patches when the clothing wore out in spots. Other pieces went into patchwork quilts.

Quilts were very personal. For some women they were the only means of creative expression, the only relief from days of toil. Women at times would spend many years on one quilt. They would get together in a quilting bee to make "presentation quilts" in honor of the visit of a famous person or a newly wedded couple. Some quilts were handed down to children and grandchildren for many generations. Others are now in museums.

Students will enjoy looking at some of the many books of quilt patterns that are now available, or you might want to arrange a visit from a local quiltmaker.

OBJECTIVES FOR STUDENTS

◆ To work with geometric shapes and compute their areas.

◆ To analyze designs for rotational and line symmetry.

Extension Activities

1. Children can create a variety of quilt patterns. Before drawing and coloring the design, students might work with paper or plastic squares and triangles that they can move about until they are satisfied with a pattern.

2. Plan a trip to a museum that displays quilts.

3. Arrange a quilt exhibit for the corridors of your school. Call in the press to write it up. Perhaps a local museum will offer to house the exhibit.

Answers to Student Pages

Page 77:
1) Block A: 8 triangles, 4 sq. units; 5 squares,
5 sq. units; Sum: 9 sq. units.
Block B: 16 triangles, 8 sq. units; 1 square, 1 sq. unit;
sum: 7 sq. units

Page 78:
1) a. 4   b. 4;

Name: _____

⬚ SQUARES AND TRIANGLES

Patchwork quilts made up of different geometric shapes are very much a part of early American culture. It was common practice for colonial women to gather and sew beautiful quilts made from different patterns which they'd invent.

Use the two different 3 x 3 quilt blocks below to answer the following questions.

Block A has ____ triangles. Their total area is _____ sq. units.
Block A has _____ squares. Their total area is _____ sq. units.
The sum of the areas in Block A is _____ sq. units.

Block B has ____ triangles. Their total area is _____ sq. units.
Block B has _____ squares. Their total area is _____ sq. units.
The sum of the areas in Block B is _____ sq. units.

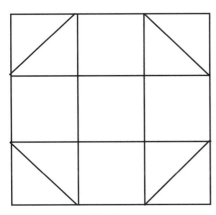

Block A

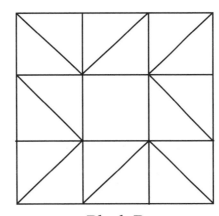

Block B

Now Try This

☞ Plan with your group to make a paper patchwork quilt. Each person will make two or three blocks. Then fit them all together.

Name: _____

 # MAKING A STAR

This 3 x 3 quilt block is called "Ohio Star." The quiltmaker might use the same pattern for all the blocks in her quilt, or she might vary it with other designs. She thinks about the *symmetry* of each block and how it will look next to others of the same or a different design.

1. a) Analyze the 3 x 3 "Ohio Star" quilt block. Rotate this sheet slowly until the pattern looks the same as the original. In how many different positions does the design look the same? b) Now imagine that you fold the block in half so that the two halves of the pattern match. How many different fold lines can you find? Draw diagrams to show them.

Here is one diagram; the dotted line is the fold:

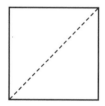

2. On a separate sheet, draw a large 3 x 3 block and design a geometric quilt pattern using two colors. Write a description of your pattern and give it to your partner. She or he will read your description and try to draw your pattern. You will do the same with your partner's pattern. Make it interesting and not too easy!

Now Try This ━━━━━━━━━━━━━━━

☞ Plan a patchwork quilt made from material. Will each block have the same pattern or will you vary the blocks? Draw and color the patterns you will need, using the correct dimensions.

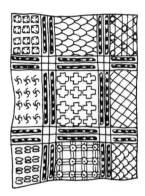

ADINKRA CLOTH OF GHANA

Background

The Asante (also spelled Ashanti) people live in Ghana, in West Africa. They make special kinds of cloth. One kind is called *kente*. Another type of cloth is *adinkra*, a word that means "goodbye."

Originally worn at funerals to honor the memory of the dead, the adinkra cloth now comes in a variety of colors and is worn for many other important occasions.

Getting Started

Students should locate Ghana on the map. The Asante belong to the Akan group of peoples, one of several groups living in Ghana.

Before starting the lesson, show students photographs of adinkra cloth—or the real thing, if it is available.

Introduce *symmetry, line symmetry,* and *rotational symmetry.* You might enlist the use of mirrors to teach this concept. Line symmetry can be tested by matching the reflection of half the design with the original half. Start by analyzing symmetrical letters (A, C, H, M, W, etc.), then move on to symbols. Students may want to analyze more familiar symbols first. Help students understand that a hand-made design may not have the near-perfect symmetry of a factory-made product. Even human beings are not exactly symmetrical! They should be ready to make allowances for such irregularities.

OBJECTIVES FOR STUDENTS

♦ To subdivide a rectangle into congruent squares.

♦ To analyze several motifs for reflectional and rotational symmetry.

♦ To learn about the adinkra cloth of the Asante people.

For an accurate analysis of rotational symmetry, students can use the "trace and turn" test. Trace the design on a sheet of clear plastic or thin paper and place it directly over the original. Mark one particular point (not the center) on the original and on the tracing. Stick a pin or a pencil point into the center so that it stays fixed as the tracing is rotated. Count the number of positions in which the tracing coincides with the original before it comes back to the starting position. Ask students how many degrees it has turned each time.

Extension Activities

1. Discuss the symbolism of the heart. Can students name symbols in our culture that have specific meanings? They may name the heart, of course, along with religious symbols, the peace symbol, traffic signs, manufacturer's logos, etc.

2. Students can create their own adinkra, either on paper or on cloth. Each student might make a "stamp" by cutting a design into the surface of a half potato or a sponge. Dip the stamp into finger paint and repeat the pattern on a rectangular paper or cloth, then place a border around it. The rectangles can be combined by pasting or stitching to make a large display.

3. Investigate the Yoruba (Nigeria) *adire* painted cloth.

Answers to Student Pages

Page 81:
The rectangle has 2 rows, each row has 5 hearts, a total of 10.

Page 82:
Note: If students have not encountered infinity before, pattern F offers a good opportunity for them to learn by experience. Some may need guidance as they keep struggling to count the positions.

	Number of Axes	Order of Rotation
A	1	1
B	2	2
C	0	4
D	2	2
E	0	2
F	Infinite Number	Infinite Number
G	2	2
H	0	1

Name: _____

A PATTERN OF HEARTS

The adinkra cloth of the people of Ghana is large and consists of many rectangles sewed together. Each rectangle is stamped with rows of a pattern, different in each rectangle. Each pattern has a meaning. Here is one rectangle. The pattern shows the talons of an eagle.

The heart pattern is called Akoma. It means "take heart" or "have patience." Complete the rectangle below with a repeated pattern of hearts.

HEART
"Have patience"

The rectangle has _____ rows.
Each row has _____ hearts.
All together the rectangle has _____ hearts.

Now Try This

☞ Make your own adinkra cloth on the back of this sheet. Start by drawing a large rectangle, then making up a pattern and repeating it. You might want to first divide the rectangle into squares to guide you. When you're done, write down the name and the meaning of your pattern. Finish by coloring your adinkra cloth.

Name _____
Meaning _____ [symbol]

Name: _____

FINDING SYMMETRY

Below are some symbols and ideas that are represented in adinkra cloths. The symbols have different kinds of symmetry. Some have line symmetry, some have turn (rotational) symmetry, and some have both.

1. Line Symmetry. Can you fold the design so that one half exactly fits over the other half? Draw that fold line. Does the design have another fold line? Draw all the fold lines you can find. Each fold line is called an axis. Write the number in the table.

Symmetry Chart

2. Rotational Symmetry. Turn the paper until the design looks the same as the original. Keep turning. In how many different positions does the design look the same? Include the starting position. This number is called the order. Write the number in the table.

	Number of Axes	Order of Rotation
A	1	1
B		
C		
D		
E		
F		
G		
H		

A

B

C

D

E

F

G

H

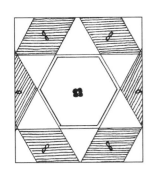

TESSELLATIONS IN ISLAMIC CULTURE

Background

The religion of Islam developed in the Middle East in the seventh century and spread into northern Africa, Spain, and Turkey, as well as eastward in Asia. Islamic artists and architects constructed beautifully decorated mosques and palaces. The predominant motifs in art were geometric forms, floral designs, and calligraphy, particularly quotations from the Koran. Geometric patterns were executed with just the compass and straight-edge. Some sects prohibited the depiction of human and animal forms.

OBJECTIVES FOR STUDENTS

♦ To work with geometric shapes and angle measurements.

♦ To create geometric tessellations.

Getting Started

Have students locate Iran on the map, as well as some of the countries that are now Islamic. Show the students photographs of Islamic buildings and decorative art (see the Metropolitan Museum of Art reference on page 118). Note the prominence of the three design elements. Some artists did depict human beings, as in the various versions of the *Shah-nameh*, the Book of Kings of Persia (now Iran), with their breathtaking miniature illustrations.

This lesson calls for the use of pattern blocks. If they are not available and cannot be ordered, perhaps you can purchase inexpensive pattern block paper shapes, which correspond in shape and color to pattern blocks. Pattern block templates are useful for recording patterns, as is isometric (triangular) grid paper.

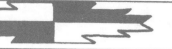

Extension Activities

1. Ask students to identify some repeated patterns in their environment.

2. Students can construct a tile panel by cutting out and decorating many congruent hexagons and attaching them to a large sheet of oak tag.

3. Students might investigate and write about tessellations in their environment. Brick walls and tile floors are obvious examples.

Answers to Student Pages

Page 85:
1) Triangle, parallelogram (or rhombus or diamond), hexagon. Students may observe that the triangles and hexagons are regular—their sides are equal in length and their angles are equal in measure. The three different shapes have matching sides.
2) All the pattern blocks tessellate: triangles, parallelograms, squares, hexagons, trapezoids.

Page 86:
1) The sides all have equal length.
2) 360°; 60°; 120°; 60° and 120°.

Name: _____

 # REPEATING PATTERNS

Look at the pattern on this page. It is formed with tiles of different shapes. It decorated the palace wall of a Persian king many centuries ago.

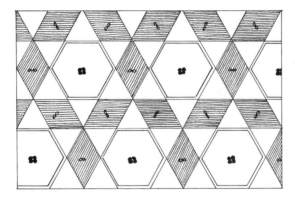

1. On a separate sheet, list the geometrical shapes that you see in the pattern above.

The shapes in the pattern fit together so that there are no empty spaces. The same design is repeated over and over again. This type of repeated pattern is called a *tessellation*.

2. Choose one pattern block shape at a time to make a tessellation. Can you make a tessellation with just triangles? _____ Parallelograms? _____

Name other pattern blocks that tessellate.

Now Try This ─────────────────────

☞ On a separate sheet, make a tessellation using two different pattern block shapes. Name the shapes you will use:

Trace your pattern block tessellation and color it. Islamic artists liked to paint flowers or other designs on some of the tiles.

Name: _____

FINDING MORE PATTERNS

Many Islamic artists used just three types of motifs: geometry shapes, flowers, and calligraphy (writing). Some Islamic groups believed it was not proper to portray humans and animals.

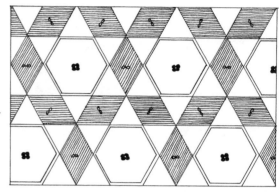

Geometric tessellations were favorite ways of decorating the walls and floors of important buildings, like mosques and palaces. Artists had to plan carefully so that the various shapes fit together with no empty spaces. They started with one shape, then spread outward in all directions from that shape.

1. Examine the pattern. How do the sides of the three different shapes compare? _____

**2. Look at one point where several shapes meet.
What is the sum of the angles around that point?** _____.

Write the measure of each angle of the triangle: _____;
the hexagon _____;
the rhombus (two different measures): _____.

Now Try This ━━━━━━━━━━━━━━━━━━━━━━━━

☞ Using pattern blocks, copy the tessellation on this page. Which shape will you use to start? Trace your tessellation on the back of this page, and color it. You might want to decorate your tiles with flower designs or writing.

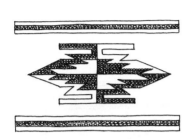

NATIVE AMERICAN BEADWORK

Background

Europeans gave the name "Sioux" to the Dakota people. Dakota means "society of friends" and signifies a confederation of allies. By the early eighteenth century they were roaming the plains on horseback to hunt buffalo. Buffalo skins were used to make tipis, clothing, and many other objects, and were often decorated with beadwork or painting.

Getting Started

Native peoples used a variety of materials to make beads: shells, bones, claws, stones, and minerals. The shell beads of the East Coast tribes were later used as currency called wampum (see page 20). Early beads, named pony beads, were supplanted in the mid-1800s by smaller, round beads for sewn beadwork.

Show students examples of Native American beadwork. Ask them how they would calculate the number of beads required for any specific pattern. Let them discuss the subject before they begin to work on the lesson. Students will need grid paper for activities in both reproducibles.

Extension Activities

1. There is evidence that the ancient Egyptians first made sketches on a grid before transferring the designs to the walls of temples. By drawing a larger grid on the wall, they were able to maintain the proper proportions in the figures they drew. Students might research this topic.

> ### OBJECTIVES FOR STUDENTS
>
> ◆ To learn about Native American beadwork patterns.
>
> ◆ To reproduce a design in a different size.
>
> ◆ To compare the areas of similar designs.

87

2. Students might be challenged to do beadwork. Consult Hunt (1954) or another book on Native American crafts.

3. Ask students to recreate the flower on page 90 *three* times the size of the original. Ask them to discuss with their groups how they would calculate the number of beads they will need.

Answers to Student Pages

Page 89:
1) Ten squares; 100 beads.
3) Forty squares; 400 beads; four times as many.

> ### COUNTING BY SYMMETRY
> Students can count the partial squares by noting the symmetry of the design.

Page 90:
1) 4 1/4 sq cm; 128 beads. Students should discuss how they counted the partial squares, and how they knew that four small squares made one square centimeter.
2) 17 sq cm; about 510 beads.
3) One-fourth, since the smaller design has one-half the length and width of the larger.

Name: _____

 # DESIGNING A TIPI

Josie is planning to decorate her jacket with beadwork in the style of the Sioux Indians. She finds a pattern for making a tipi. Now she must figure out how many beads she will need. Can you help her?

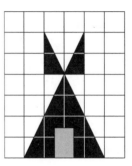

1. The pattern is printed on grid paper. About ten beads will cover one small square. How many squares must be covered? _____ squares. How many beads will she need for the pattern? _____ beads. Describe how you figured it out.

Eddie likes the tipi pattern. He wants to decorate his jacket with beadwork, too. But he wants his tipi to be twice as high and twice as wide as the tipi in the grid pattern. He draws his design on grid paper. Then he counts the squares and figures out how many beads he will need.

2. Draw Eddie's tipi on the grid. First outline it lightly in pencil. Be sure that it is two times as high and two times as wide as Josie's. Eddie's design is similar to Josie's. It has the same shape but a different size.

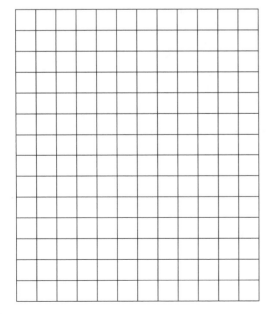

3. Count the small squares in Eddie's design. Figure out how many beads he will need. The design covers _____ squares. Eddie needs about _____ beads. He needs _____ times as many beads as Josie.

4. On a separate sheet of grid paper draw a tipi that is four times as high and four times as wide as Josie's. Color your tipi.

Name: _____

 # DESIGNING A FLOWER

The Native people of North America made beautiful beadwork designs on clothing, bags, and other items. Different nations developed their own styles. People in the Great Lakes region often made flower designs, like the design on the grid. Each small square of the grid measures 1/2 cm.

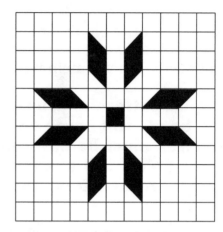

   Suppose you want to make a beadwork pattern like this one. You must estimate the number of beads you will need. Will you copy the pattern exactly, or will you make it bigger or smaller?

1. On a sheet of grid paper copy the pattern exactly. About 30 beads can cover one square centimeter.

The area of the design is _____ sq cm. The design requires about _____ beads. Describe how you worked it out.

2. Suppose you decided to make a design that was similar to this one. You want the flower to have the same shape, but each part should have twice the length and width of the original design. Draw this pattern.

It covers _____ sq cm and requires about _____ beads.

3. Draw a geometric pattern on grid paper. Now make a similar design that has half the width.

The area of the small design is _____ the area of the large design.

Part IV

FUN AND GAMES

Fun and games" is the theme of the last six lessons, introducing games, puzzles, and activities from many cultures.

River-crossing puzzles are centuries old, and probably arose from practical transportation problems. The versions described in Lesson 20 are from the Sea Islands of South Carolina and from Liberia. They offer excellent opportunities for dramatic play in the course of problem-solving.

Lesson 21 explains both a simplified and an authentic version of Oware, played in Ghana and Nigeria. This game, under various names and with different rules of play, is popular throughout Africa and in other regions as well. Students develop a feeling for numbers while they apply problem-solving strategies to figure out the best moves.

In Lesson 22 students learn Shisima, a three-in-a-row game from western Kenya, and Mu Torere, a three-in-a-corner game from New Zealand. Besides developing strategies for winning, students use their skills in geometry and measurement to draw the eight-pointed gameboards. Lesson 23 presents two additional three-in-a-row games: Tapatan from the Philippines, and Picaría, played by Pueblo children in the Southwest of the United States. Similar gameboards were incised in the roofing slabs of an Egyptian temple thirty-three centuries ago.

In Lesson 24 students practice their computational skills by assigning numerical values to the letters of the alphabet, in the tradition of the ancient Hebrews and Greeks.

Probability is the theme of Lesson 25. Students spin a six-sided top to play the Mexican game Toma Todo, and toss cowrie (or macaroni) shells for the Nigerian game Igba-Ita. They compare the outcomes when an asymmetric shell is thrown with those for a symmetric coin, and learn about the basis for predicting outcomes.

Book Links for Students

Bell, Robbie & Michael Cornelius. *Board Games Round the World.* New York: Cambridge University Press, 1988.

Burns, Marilyn. *The $1.00 Word Riddle Book.* New York: Cuisenaire, 1990.

Corwin, Judith Hoffman. *African Crafts.* New York: Franklin Watts. 1990.

Orlando, Louise. *The Multicultural Game Book.* New York: Scholastic, 1993.

Zaslavsky, Claudia. *Tic-Tac-Toe and Other Three-in-a-Row Games.* New York, Crowell, 1982: 33–37.

CROSSING THE RIVER

Background

This puzzle has many versions; the simplest are included in this lesson. Other versions involve three jealous husbands and their wives, in-laws who fear a quarrel, masters and servants, merchants and robbers. In each version certain characters cannot be left alone with certain other characters. The version about the two animals and the food has been traced back to the eighth century teacher Alcuin of York, who wrote about these puzzles in a letter to Charlemagne, his most famous student.

Getting Started

This lesson requires students to work in groups of four. If at all possible, they should have room to act out their versions of the river crossing. If there is insufficient space in the classroom for all groups to do this simultaneously, organize the groups so that they take turns.

Extension Activities

1. Students might experiment with altering the conditions of the puzzle; e.g., the man must transport four items that are mutually hostile in certain combinations.

2. Ask students to make a comic strip showing how the man was able to cross the river with the leopard, the goat, and the bundle of cassava leaves.

OBJECTIVES FOR STUDENTS

♦ To learn an ancient puzzle in versions from the Sea Islands of South Carolina and from Liberia.

♦ To apply logical reasoning to the solution of the puzzle.

3. When discussing Fox, Duck, and Corn, ask students: How many times did the man cross the river? (three) How many round trips did he make? (one and a half)

4. When discussing Leopard, Goat, and Cassava, ask students: How many times did the man cross the river? (seven) How many round trips did he make? (three and a half)

Answers to Student Pages

Page 95:
1)

Start		Finish
Man Duck Fox Corn	⬭	
Duck	(Man / Fox / Corn) ➡	
Duck	⬅ (Man)	Fox Corn
	(Man / Duck) ➡	Fox Corn
		Man Duck Fox Corn

An alternative is for the man to transport the duck first, then go back for the fox and the corn. Why are these the only solutions that do not involve unnecessary trips?

Page 96:
1) In this version of the puzzle, the man can transport only one item besides himself. Students should see the difference between this story and the story in Part 1.
This is one solution:

An alternative solution interchanges the leopard and the cassava.

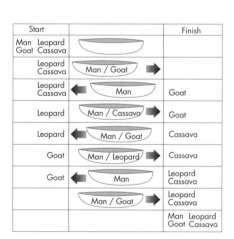

Start		Finish
Man Leopard Goat Cassava	⬭	
Leopard Cassava	(Man / Goat) ➡	
Leopard Cassava	⬅ (Man)	Goat
Leopard	(Man / Cassava) ➡	Goat
Leopard	⬅ (Man / Goat)	Cassava
Goat	(Man / Leopard) ➡	Cassava
Goat	⬅ (Man)	Leopard Cassava
	(Man / Goat) ➡	Leopard Cassava
		Man Leopard Goat Cassava

Name: _____

 # Fox, Duck, and Corn

People all over the world tell this puzzle. They may change the names of the animals and the food. That doesn't change the way to solve the puzzle. African Americans who live in the Sea Islands of South Carolina may have brought this story from Africa. They changed the names of the animals and the food to those familiar in America.

A man has a fox, a duck, and a bag of corn. He has to take them across the river in a small rowboat. The boat can hold only the man himself and two other objects. The man cannot leave the fox with the duck because the fox loves duck meat. He cannot leave the duck with the corn because the duck will eat it. How does the man get the fox, the duck, and the corn to the other side of the river?

1. **In your group, talk about one way that the man can carry all three things across. He may have to make more than two trips. Write your plan so that you don't forget it.**

   Be prepared to act out your plan. Each student will play the part of a different character in the story. Make a sign for each student with the name of the character. Does your plan work? If not, try another plan. Can you find more than one way to solve the problem?

2. **Work with your group to draw pictures of the man and his three things. Each person can draw a different picture. Decide what each person will draw. Color the pictures. Write the story about the pictures. Put everything together to form a little book. Make a cover for your book.**

Name: _____

 # LEOPARD, GOAT, AND CASSAVA

"Crossing the river" puzzles are well over a thousand years old. As they traveled around the world, people would change the characters to fit their own culture. Here is the way they tell the story in Liberia, a country in West Africa.

A man has a leopard, a goat, and a pile of cassava leaves. He must take them across the river in a boat. The boat can carry no more than one at a time, besides the man himself. He cannot leave the goat with the cassava leaves because the goat will eat them. He cannot leave the leopard with the goat because the leopard will eat it. How will he get all three—the leopard, the goat, and the bundle of cassava leaves—across the river?

Each student in your group plays the part of a different character in the story. Yes, a bundle of cassava leaves is also a character. Decide how you will know which character each person plays.

Plan one way to solve the puzzle. Write out your plan on the back of this sheet. Act it out and see whether it works. If not, plan another way to solve the puzzle. Write out the plan that works. Is there more than one way to solve the puzzle?

THE OWARE GAME

Background

Oware is a version of one of the world's oldest and best-known games. Known by many names—Mancala in Arabic, Bao in Swahili (in East Africa), Sungka (in the Philippines)—it is based entirely on mathematical principles. The rules and the shape of the gameboard may vary from one region to another. West Africans and Filipinos favor a two-row board, while in eastern and southern Africa the four-row board is more popular.

The Asante people of Ghana play several versions of Oware. The one given here is identical to a version of the Ayo game of the Yoruba (southwest Nigeria). The terms used in the game reflect the culture of the people. In southern Ghana, players of Adi "buy houses" and place their wealth in the "treasury". Some people call the playing pieces "cattle," while others "take prisoners," "sow seeds," or "eat" the opponent's beans.

Getting Started

The reproducibles offer a simplified and full version of the same game. You may use the simplified version as a stepping stone to the full version. You will find it helpful to play the game yourself in advance so that you become familiar with its moves and strategies.

Make sure students have partners to play with. If there is an odd number of students, you could be the extra partner.

OBJECTIVES FOR STUDENTS

◆ To learn an ancient game played in Africa, Asia, and America.

◆ To develop a numerical sense.

◆ To apply logic and higher order thinking skills in a game of strategy.

Extension Activities

1. Invite students to change the rules of the game—as long as both players agree! They might start with five beans in a cup, or vary the conditions for capture. One version played in Kenya has the players alternating the direction of play on each move. (See Zavlavsky, 1979.)

2. Students might decorate their gameboards and stage a cultural festival with other appropriate materials.

Answers to Student Pages

Page 100:
To simplify the game somewhat as students learn it, they may start with just three beans in each cup—36 in all.

The rules for capturing the opponent's beans may be confusing to the students. It might be helpful for you to draw a "gameboard" on a transparency and set up several situations that students can discuss. The diagrams below illustrate one situation.

Player One is about to move. She picks up the five beans in cup D and sows them one by one in the five cups to the right: E, F, I, II, and III. She captures the three beans in cup III, the two beans in cup II, and the three beans in cup I, because they are in an unbroken sequence on the opponent's side of the board. She places these eight beans in her endpot. She cannot capture the three beans in cup F because that cup is on her side of the board.

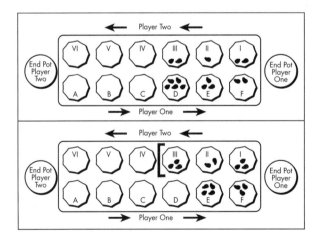

98

Name: _____

 # LEARNING THE RULES

You and your friend decide to play the Oware (oh-WAHR-ee) game. Your friend has never played the game. To help your friend learn the rules, you will make the game simple at first.

You will play Oware with:

- A sheet of paper divided into eight equal parts. Call it the "board."
- Sixteen large beans or counters.
- Two small bowls, called "endpots," to hold the captured beans.

To start: The players face each other, with the board between them. Place two beans in each space, called a "cup." Each player has the four cups on the near side and the endpot to her or his *right*.

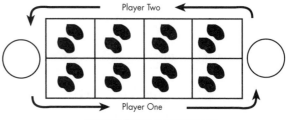

STARTING POSITION

To move: Player One picks up all the beans in one of her four cups. She drops one bean in each cup going around the board to the right.

Player Two picks up all the beans in one of his cups. He drops one bean in each cup going around the board to the right.

To Capture: A player captures from the *far* (opponent's) side of the board. If the last bean in any move makes a group of two in a cup on the far side, these beans are placed in the player's endpot. Then, going backward, if the cup next to this one also has two beans, these are captured and placed in the endpot. Continue with the cup just before this one, as long as it is on the far side of the board and has two beans.

To Finish: The game ends when one person has no beans left on her or his side of the board.

Name: _____

 # HOW IT'S PLAYED IN GHANA

Oware is one form of a game that is played in most of Africa and parts of Asia. Many years ago African captives brought the game to America. The game has many names and sets of rules. People in West Africa play this version. *Oware* is one name of the game in Ghana.

Getting Started: Some game experts think that this game is among the ten best in the world. It is played with cowrie shells on beautifully carved boards, or with pebbles in two rows of holes dug in the dirt.

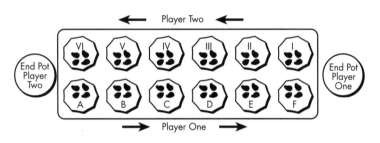

This game is for two players. They sit facing each other, with the board between them.

You will play Oware with:
- An egg carton with twelve cups, called the "board."
- 48 large beans or counters.
- Two small bowls, called "endpots," to hold each player's captured beans.

To Start: Place four beans in each cup. Each player has the six cups on the near side and the endpot to his or her *right*.

To Move: Player One picks up all the beans in any one of her six cups. She drops one bean in each cup going around the board to the right until she has dropped all the beans in her hand. Player Two does the same, starting on his side of the board. They continue, taking turns.

To Capture: When the last bean dropped in a cup on the *far* side of the board makes a group of two or three, those beans are captured; the same with the cups just before this one on the far side of the board.

To Finish: When one player has no beans left on his side, the other player must move so that she gives him some beans, if possible. If only one or two beans remain on the board, they go to the player whose side they are on. The player who has captured more beans is the winner.

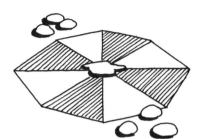

BOARD GAMES

Background

Three-in-a-row gameboards have been traced back to the diagrams incised in the roofing slabs of the 3300-year-old temple to the Egyptian pharaoh Seti I. One diagram is amazingly like the Shisima board.

Getting Started

Make sure students have partners to work and play with. Note that each pair will need two sets of counters of different colors (sets of three for Shisima, four for Mu Torere). They will need a large sheet of stiff paper to make the gameboard.

OBJECTIVES FOR STUDENTS

◆ To learn games from Kenya and New Zealand.

◆ To use geometry and measurement concepts to make gameboards.

◆ To devise strategies in games of skill.

Extension Activities

1. Students might research the cultures and histories of the peoples who play Shisima and Mu Torere.

2. Drawing the gameboard is a good application of geometry and measurement. Encourage students to make game boards of different sizes and material. Challenge students to make a "life-size" board where the students are the actual playing pieces.

3. Encourage students to learn other foreign words from Kenya or New Zealand.

4. Encourage students to discuss the strategies they used when playing Mu Torere. Here are some questions you might ask them: Why are certain counters not permitted to move into the putahi at the beginning of the game? (Students should try such a move and realize that the opponent can win in two moves.) Why does the winning move require a right-angle formation with an empty point between the two outer counters? (Again, students should note the position of the counters, as in the previous discussion, and see that the opponent is indeed blocked from moving.)

TEACHER TIP

It is a good idea to have students learn Shisima before they tackle Mu Torere. The moves in both games are similar, but Shisima is easier to learn.

Name: _____

 # SHISIMA FROM KENYA

Children in Western Kenya play a game called *Shisima*, which means "a body of water." The counters are called *imbalavali*, or "water bugs." These bugs move so fast that it is hard to follow them with your eyes. That's how fast some children play Shisima.

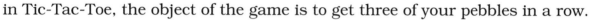

The game is something like Tic-Tac-Toe. Two people play on a "gameboard" drawn in the sand. They use two kinds of pebbles for counters. As in Tic-Tac-Toe, the object of the game is to get three of your pebbles in a row.

You will play Shisima with:

- A gameboard in the shape of an octagon (eight-sided polygon).
- Three counters of one color (white) and three of another color (black).

Discuss how you will draw the gameboard on a large sheet of paper. Start with a circle. Mark the center, then draw four diameters, as in the diagram. Connect the endpoints of the diameters. In the center draw a *shisima*, a body of water. The game is played on the eight points of the octagon and on the shisima in the center.

To Start: Place the counters on the gameboard as in the diagram. In this round, Player One has white counters and Player Two has black counters. They can change colors for the next round.

To Move: Player One moves a white counter one space along a line to the next empty point. Then Player Two moves a black counter one space along a line to an empty point. A player may move into the center (the shisima). Jumping over a counter is not allowed.

Object: To place your three counters in a straight line.

To Finish: The first player to place the three *imbalavali* in a row is the winner. The game ends in a tie when the same set of moves has been repeated three times.

Name: _____

MU TORERE FROM NEW ZEALAND

In New Zealand, Maori children—and adults, too—play Mu Torere. Two people play on a gameboard in the shape of an eight-pointed star. The object of the game is to block your opponent from moving.

You will play Mu Torere with:

Mu Torere
Gameboard

- A gameboard in the shape of an eight-pointed star, with a center space called a *putahi*, the place where paths or streams run into one another.
- Four counters of one color (black) and four of another color (white).

Decide how you will draw the gameboard on a large sheet of paper. You might start with a circle. Draw four diameters to divide the circle into eight equal sectors (like a pie). Then draw the star and the center *putahi*.

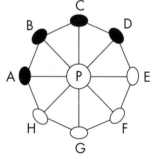

Starting Position

To Start: Place the counters as in the diagram. Black goes first.

To Move: On the first two moves for each player, the counters in the outer positions—points A, D, E, and H—may not move into the center. Otherwise the players take turns moving one counter at a time one space to an empty point or empty *putahi*. Jumping over a counter is not allowed.

Object: To prevent the other player from moving.

To Finish: The only way to block your opponent is to place three of your counters in a right-angle formation to make a "corner," with an empty point between the two counters on the points of the star.

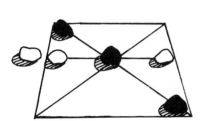

THREE-IN-A-ROW GAMES

Background

Tapatan is another name for the popular game known as Three Men's Morris or Mill. Versions of the game are found in most parts of the world. Many gamebooks include more complex three-in-a-row games, such as Nine Men's Morris.

Stewart Culin, in his book *Games of the North American Indians* (Dover, 1975), describes his conversations with Pueblo youngsters about the game they called Picaría or Pitarilla, names that are probably based on the Spanish word for "little stone." Although the Pueblo were terribly oppressed by the Spanish conquistadors, they adopted Spanish three-in-a-row games, and perhaps changed them to their own versions.

Getting Started

Students will need partners for both reproducibles. Each student pair will need two large sheets of paper for making game boards and two different-colored sets of three counters.

Extension Activities

1. Students can investigate other three-in-a-row games, such as Nine Men's Morris (see Orlando, 1993; Krause; Seattle Public Schools; Zaslavsky [1979 and 1982]).

OBJECTIVES FOR STUDENTS

♦ To learn games of the Philippines and of the Pueblo of New Mexico.

♦ To use geometry and measurement to draw boards for games.

♦ To devise strategies in games of skill.

♦ To compare two similar games.

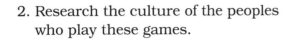

2. Research the culture of the peoples who play these games.

3. Invite students to discuss the geometry of the game board square and terms like "diagonal" and "midpoint." Students should draw accurate figures. Discuss how to draw the square so that the sides are equal in length and the angles are right (90°) angles. Students may want to fold paper in appropriate ways.

4. Encourage students to make up variations of the game.

5. Discuss the different strategies to Picaría. Ask why a player may not place a counter in the center in the opening stage. What happens if she does? Of the 16 different ways to make a row, how many go through the center? (8).

VARIATIONS OF TAPATAN

In the French version of Tapatan called Marelle, neither player may use the center point for the first move. A version of Achi, played in Ghana, calls for each player to use four counters. Tant Fant, an Indian (Asia) game, opens with each player's three counters already in position along the outer horizontal rows closest to the players as they face each other.

Answers to Student Pages

Page 108:
Five squares lie within the large square. Students should recognize the tilted square formed by the midpoints of the large square. There are 13 points where lines meet or intersect.

Some of the factors to be considered when comparing the two versions are:
- Is the first person to move more or less likely to win?
- How long is each game?
- Which version is more challenging?

Name: _____

 # TAPATAN FROM THE PHILIPPINES

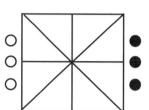

Children and grownups in the Philippines play Tapatan, a game much like Tic-Tac-Toe. It is played on a game board with nine points. The two players move their counters on the board. Each player tries to make a row of three counters.

Three ways to place three counters in a row:

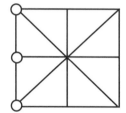 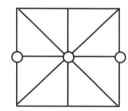 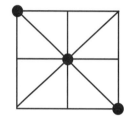

You will play Tapatan with:

- A game board in the shape of a square.
- Three counters of one color (white) and three of another color (black).

Discuss how you will draw the game board on a large sheet of paper. How will you know that the board is a square? How many more lines must you draw? The game is played on the nine points where the lines meet.

To Start: Player One puts one white counter on any point on the board. Player Two puts one black counter on any empty point on the board. They take turns placing their counters on empty points on the board until all six counters are on the game board.

To Move: Player One moves one white counter one space along a line to the next empty point. Player Two does the same with a black counter. Jumping over a counter is not allowed.

Object: To place your three counters in a straight line.

To Finish: The winner is the first player to make a row of three. The game ends in a tie when the same set of moves has been repeated three times.

Name: _____

 # PICARIA FROM THE PUEBLO

A hundred years ago and more, Pueblo children of New Mexico were playing a three-in-a-row game that they called Picaría. The rules might differ from one village to another. Here is one set of rules.

You will play Picaría with:
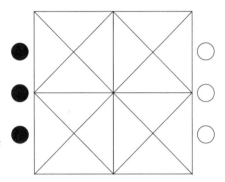
- A square game board marked as in the diagram.
- Three counters of one color (black) and three of another color (white).

Work with your partner to draw the game board on a large sheet of paper. How many squares do you see inside the large square? The game is played on all the points where lines meet. How many such points are there?

To Start: Player One has black counters. The two players take turns placing their counters one at a time on any point, except for the center point.

To Move: Player One moves one black counter one space along a line to the next empty point. Player Two does the same with a white counter. Now they may occupy the center point. Jumping over a counter is not permitted.

Object: To place your three counters in a straight line, with no empty points between your counters. There are 16 different ways to make a row.

To Finish: The winner is the first person to make a row of three.

Variation: Here is another way to play Picaría. Just change the game board a bit. Play on the nine points that are marked with black dots. Now there are only eight ways to place three counters in a row.

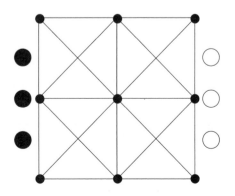

Discuss with your partner the two different ways to play Picaría. Make lists of the good points and the bad points of each version. What factors will you consider?

GEMATRIA: NUMBERS AND LETTERS

Background

Number mysticism and the association of numbers with various qualities go far back in history, and are identified with Pythagoras and his school. They regarded even numbers as feminine and odd numbers as masculine.

The ancient Hebrews made substitutions in their numerals to circumvent a taboo. According to their system of using letters as numerals, the letters for the number fifteen would be *yod heh*, ten and five. But this combination represented the spelling of the forbidden name Yahweh (Jehovah). Therefore they substituted the letters for nine and six.

Martin Stifel, a prominent German algebraist of the sixteenth century, was an ardent follower of the religious reformer Martin Luther. Using Gematria, he "proved" that Pope Leo X was the "beast" in the Bible because the value of his name was 666. Later it was Martin Luther who was considered the beast, on the basis of similar reasoning.

You might ask students to use mental arithmetic for easy calculations, and calculators when the going gets rough.

OBJECTIVES FOR STUDENTS

♦ To learn some numerical practices of ancient Greeks, Jews, and Christians.

♦ To solve problems involving the operations of addition and subtraction.

♦ To use logic to solve problems.

Extension Activities

1. Ask students to research the history of gematria. The topic may be called "numerology" or "arithmography". (See Ifrah, 1985.)

2. Look for $1.00 words, using the system in Part 1 and counting the value of each number in cents. (See Burns 1990; Whitin & Wilde, 1992: pages 200–201).

3. When discussing "Math with Letters" (page 112) ask students to describe how they knew which letters to use in #3. The answer: Find the value of the given letters and subtract from the sum. If only one letter is missing, fill in that letter. When two letters are missing, try various combinations of two letters that give the required number. It may be obvious that one letter is a multiple of ten and the other is less than ten.

Answers to Student Pages

Page 111:
1) Answers will vary.
2) CAT, EAR, BELT, BUG, CAMP or COMB.
3) Answers will vary.
4) For example: AT, OF, BAR, PAD, LED, HID, LACE, DICE.

Page 112:
1) JD, KI, L, MA, NE.
2) a. NE   b. MF   c. N   d. JD
3) HE, BAD, FALL, BAND, MADE or DAME.

Name: _____

 # TURNING WORDS INTO NUMBERS

What number does your name add up to? Tom found that his name adds up to 48. Ellen's name also adds to 48. They are surprised to find that both names have the same value. ELLEN has five letters, while TOM has only three.

1. Write the value of each letter in the chart, using numbers from 1 to 26. Then add the values of the letters in your name.

A	B	C	D	E	F	G	H	I	J	K	L	M
1	2	3	___	___	___	___	___	___	___	___	___	13

N	O	P	Q	R	S	T	U	V	W	X	Y	Z
___	___	___	___	___	___	20	___	___	___	___	___	26

My name:
(letters) _____ = **(numbers)** _____

2. Write the missing letter in each word to get the sum:

C _ T = 24   _ A R = 24   _ E L T = 39

B _ G = 30   C _ M _ = 33

3. On a separate sheet write five words. Figure out the value of each word. Then rewrite the words and their values below, but leave out one or two letters. Ask your partner to fill in the missing letters. Check your partner's answers.

4. Work with a partner. Write as many words as you can that add up to 21.

Name: _____

 # MATH WITH LETTERS

When you write Roman numerals you use letters of the Latin alphabet: for example, X = 10, V = 5, I = 1. For 30 you write X three times: XXX.

The ancient Greeks and Hebrews also used letters of their alphabets to write their numerals. Here is an example, using the Latin alphabet:

A	B	C	D	E	F	G	H	I	J	K	L	M	N	O
1	2	3	4	5	6	7	8	9	10	20	30	40	50	60

Thus 23 would be KC and 68 would be OH.

1. Use the letter system to write:

14 = _____; 29 = _____; 30 = _____; 41 = _____; 55 = _____.

2. Carry out the additions and subtractions in letters, and check by changing to the numerals we use:

a. KA + LD = _____ b. NG - JA = _____

c. LE + JE = _____ d. MB - KH = _____

3. Write the missing letter or letters in each word to get the sum. Use only the letters A through O:

H __ = 13   __ AD = 7   __ A L L = 67

B A __ __ = 57   __ A __ E = 50

Now Try This ━━━━━━━━━━

☞ Make up five puzzle words like those in #3 and ask your partner to solve them. Check your partner's work.

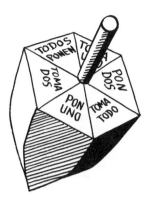

GAMES OF CHANCE

Background

One of the most important mathematical topics is the study of statistics and probability. An understanding of these ideas is essential to becoming an informed citizen. Playing games of chance is an excellent introduction to the topic. In this lesson we deal with two types of probability.

The six-sided symmetrical top in Toma Todo is equally likely to land on any one of its sides. Similarly, a fair coin is just as likely to come up head or tail when it is tossed. The geometry of the top or the coin provides the basis for making predictions about the outcomes.

We cannot say the same for the cowrie or macaroni shells in Igba-Ita. Not only is the shape of the shell irregular, but the shells differ from one another in shape. Gathering data about the actual outcomes when the shells are tossed enables us to make better-informed predictions.

> ### OBJECTIVES FOR STUDENTS
>
> ♦ To learn games of chance of Mexico and Nigeria.
>
> ♦ To collect, organize, and interpret data.
>
> ♦ To explore concepts of chance.

Getting Started

For Toma Todo you will need a six-sided spinner for each group. For Igba-Ita you will need enough macaroni shells (or half walnut shells) to provide each student with 12. Of course, real cowrie shells can be used, if available. Both games are best played with groups of four.

Extension Activities

1. Allow students to research other games of chance, such as the Dreidel game, played on the Jewish holiday Hanukkah, and several American Indian games.

2. Ask students to keep a tally of their spins as they play the Toma Todo game, Suggest that they write the six phrases on the top and put a check mark under each phrase as it comes up on a spin. Ask them to continue recording through all rounds of play. What do they notice? Students should find that, as the number of spins increases, each of the six sides comes up roughly the same number of times. Discuss why this is so. If some results diverge considerably from equality, investigate possible reasons (e.g. the spinner sections are unequal, the pencil is off-center).

MAKING YOUR TOMA TODO

You can either provide readymade spinners or have students make their own out of heavy paper. If students choose to make their own, they can follow the pattern illustrated on page 115. A pencil stub inserted through the center will serve as the spinner's axis.

3. Ask students to keep a record of their outcomes when playing Igba-Ita. At the end of their games, they can compare the outcomes for the different groups and obtain class totals. Students can use these statistics to predict the outcomes of future games.

THE SHELL GAME

Igba-Ita, which translates to "pitch and toss," is one name for a cowrie shell gambling game that is popular in many parts of Africa. The Mende people of Sierra Leone call the game *Kpoyei*. When the game is played with coins, the Igbo call it *Igba-Ego* (ego = money). The gambling game was a favorite male recreation on market day.

Name: _____

 # TOMA TODO FROM MEXICO

Children and grownups in Mexico often play Toma Todo. They use a six-sided top called a *Pirinola* or *Topa*. Probably the word *Topa* comes from the English word "top." Two or more people play the game. In this game, winning depends on luck, not on how well the people play. Will you be lucky?

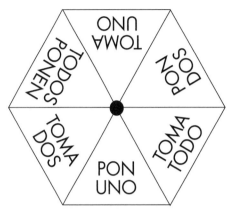

Write the Spanish words you see in the diagram. They mean:

Side	Spanish	English
1	Toma Uno	Take One
2	Toma Dos	Take Two
3	Toma Todo	Take All
4	Pon Uno	Put One
5	Pon Dos	Put Two
6	Todos Ponen	All Put

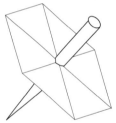

To Start: Each person should have ten chips or counters. Each player puts two chips in the center, called the "pot."

To Play: Take turns spinning the Pirinola once each. When it comes to rest, read aloud the instructions on the highest part of the top. The player may be told to take one or two or all the chips from the pot. Or the player may have to put one or two chips into the pot. "Todos Ponen" means that every player places two chips into the pot. When only one or two chips remain in the pot, every player places two chips into the pot.

To Finish: Decide before you start how many rounds you will play. A player who does not have enough chips to play drops out of the game. The winner is the person with the most chips at the end of the game.

Name: _____

Igba-Ita from Nigeria

Igba-Ita is a game of chance of the Igbo people of southeast Nigeria. In this game each player tosses four or more cowrie shells. He or she wins or loses according to the way the shells fall.

Players: Two.

You will play it with:
12 cowrie (macaroni) shells for each player.

To Play: One person (the challenger) picks up four of his or her shells. The other players agree that each will put one, two, or more shells into the center, called the "pot." The challenger tosses the shells. Players note whether they fall with the openings up or down.

To Win the Pot: The challenger wins and takes the pot when the cowries land:
(1) All four up;
(2) All four down;
(3) Two up and two down.

To Continue: If the challenger wins, he or she continues to toss. If that player loses, his or her four shells go into the pot, and the next person becomes the challenger.

To Finish: Decide in advance how many rounds to play. The winner is the person who has the most shells at the end. A person who has too few shells to play drops out of the game.

Now Try This

☞ Work with a partner. Each person tosses a shell 20 times and records the number of times it lands up or down. Compare your results. Then carry out the same experiment with coins. Compare the results with those of the shells. Write up your experiments.

TEACHER BIBLIOGRAPHY

Math and Literature Connection

Kolakowski, Jane Steffen. *Linking Math with Literature*. Greensboro, NC: Carson-Dellosa, 1992.

Seale, Doris & Beverly Slapin. *Through Indian Eyes: The Native Experience in Books for Children*. Philadelphia: New Society Publishers, 1992. Not specifically mathematical, but very useful for analyzing books for authenticity.

Thiessen, Diane & Margaret Matthias. *The Wonderful World of Mathematics*. Reston, VA: National Council of Teachers of Mathematics, 1992.

Welchmans-Tischler, Rosamond. *How to Use Children's Literature to Teach Mathematics*. Reston, VA: National Council of Teachers of Mathematics, 1992.

Whitin, David & Sandra Wilde. *Read Any Good Math Lately?* Portsmouth, NH: Heinemann, 1992.

Math and History/Culture Connection

Ascher, Marcia. *Ethnomathematics*. Belmont, CA: Brooks/Cole, 1991.

Bell, Robbie and Michael Cornelius. *Board Games Round the World*. New York: Cambridge University Press, 1988.

Bradley, Claudette. "The Four Directions Indian Beadwork Design with Logo." Arithmetic Teacher. (May, 1992): 46–49.

Caduto, Michael and Joseph Bruchac. *Keepers of the Earth: Native American Stories and Environmental Activities for Children*. Golden, CO: Fulcrum, 1988.

Carey, Deborah A. "The Patchwork Quilt." *Arithmetic Teacher*. (December, 1992): 199–203.

Closs, Michael, ed. *Native American Mathematics*. Austin: University of Texas, 1986.

Dolber, Sam. *From Computation to Recreation Around the World*. San Carlos, CA: Math Product Plus, 1980.

Grunfeld, Frederic V. *Games of the World*. New York: Ballantine, 1977.

Ifrah, Georges. *From One to Zero: A Universal History of Numbers*. New York: Viking, 1985.

Joseph, George G. *The Crest of the Peacock: Non-European Roots of Mathematics*. London: I.B. Tauris, 1991; Penguin paperback, 1992.

Katz, Victor. *A History of Mathematics*. New York: HarperCollins, 1993.

Menninger, Karl. *Number Words and Number Symbols*. Cambridge, MA: M.I.T. Press, 1969. (Paperback: Dover 1992)

Norman, Jane & Stef Stahl. "The Mathematics of Islamic Art." New York: Metropolitan Museum of Art. (Kit of 20 slides and teaching materials.)

Ohanian, Susan. *Garbage Pizza, Patchwork Quilts, and Math Magic.* New York: Freeman, 1993.

Olivastro, Dominic. *Ancient Puzzles.* New York: Bantam, 1993.

Orlando, Louise. *The Multicultural Game Book.* New York: Scholastic, 1993.

Pappas, Theoni. *The Joy of Mathematics* (1989); *More Joy of Mathematics* (1991). San Carlos, CA: Wide World Publishing/Tetra.

Van Sertima, I. *Blacks in Science.* New Brunswick, NJ: Transaction, 1983.

Whitin, David. "Looking at the World from a Multicultural Perspective." *Arithmetic Teacher.* (April, 1993): 438–441.

Zaslavsky, Claudia. *Africa Counts: Number and Pattern in African Culture.* New York: Lawrence Hill Books, 1979.

Zaslavsky, Claudia. "People Who Live in Round Houses." *Arithmetic Teacher.* (September, 1989): 18–21.

———. "Symmetry in American Folk Art." *Arithmetic Teacher.* (September, 1990): 6–12.

———. "Multicultural Mathematics Education for the Middle Grades." *Arithmetic Teacher.* (February, 1991): 8–13.

Multicultural Education—General

Allen, Judy, Earldene McNeill, & Velma Schmidt. *Cultural Awareness for Children.* Menlo Park, CA: Addison-Wesley, 1992.

Baker, Gwendolyn C. *Planning and Organizing for Multicultural Instruction* (second edition). Menlo Park, CA: Addison-Wesley, 1993.

Waldman, Carl. *Encyclopedia of Native American Tribes.* New York: Facts-on-File, 1988.

Weatherford, Jack. *Indian Givers: How the Indians of the Americas Transformed the World.* New York: Fawcett Columbine, 1988.

Multicultural Mathematics Activities

Krause, Marina. *Multicultural Mathematics Materials.* Reston, VA: National Council of Teachers of Mathematics, 1983.

Seattle Public Schools. *Multicultural Mathematics Posters and Activities.* Reston, VA: National Council of Teachers of Mathematics, 1984.

Zaslavsky, Claudia. *Multicultural Mathematics: Interdisciplinary Cooperative-Learning Activities.* Portland, ME: J. Weston Walch, 1993. (For grades 6 and up, but adaptable to lower grade level. Suitable for teacher education.)

NOTES

NOTES